Lecture Notes in Mathematics 2021

Editors:
J.-M. Morel, Cachan
B. Teissier, Paris

T0222671

For further volumes:
http://www.springer.com/series/304

Andreas Defant

Classical Summation in Commutative and Noncommutative L_p-Spaces

 Springer

Andreas Defant
Carl von Ossietzky University Oldenburg
Department of Mathematics
Carl von Ossietzky Strasse 7-9
26111 Oldenburg
Germany
defant@mathematik.uni-oldenburg.de

ISBN 978-3-642-20437-1   e-ISBN 978-3-642-20438-8
DOI 10.1007/978-3-642-20438-8
Springer Heidelberg Dordrecht London New York

Lecture Notes in Mathematics ISSN print edition: 0075-8434
                              ISSN electronic edition: 1617-9692

Library of Congress Control Number: 2011931784

Mathematics Subject Classification (2011): 46-XX; 47-XX

© Springer-Verlag Berlin Heidelberg 2011
This work is subject to copyright. All rights are reserved, whether the whole or part of the material is
concerned, specifically the rights of translation, reprinting, reuse of illustrations, recitation, broadcasting,
reproduction on microfilm or in any other way, and storage in data banks. Duplication of this publication
or parts thereof is permitted only under the provisions of the German Copyright Law of September 9,
1965, in its current version, and permission for use must always be obtained from Springer. Violations
are liable to prosecution under the German Copyright Law.
The use of general descriptive names, registered names, trademarks, etc. in this publication does not
imply, even in the absence of a specific statement, that such names are exempt from the relevant protective
laws and regulations and therefore free for general use.

Cover design: deblik, Berlin

Printed on acid-free paper

Springer is part of Springer Science+Business Media (www.springer.com)

Preface

In the theory of orthogonal series the most important coefficient test for almost everywhere convergence of orthonormal series is the fundamental theorem of Menchoff and Rademacher. It states that whenever a sequence (α_k) of coefficients satisfies the "test" $\sum_k |\alpha_k \log k|^2 < \infty$, then for every orthonormal series $\sum_k \alpha_k x_k$ in $L_2(\mu)$ we have that $\sum_k \alpha_k x_k$ converges μ-almost everywhere. The aim of this research is to develop a systematic scheme which allows us to transform important parts of the now classical theory of almost everywhere summation of general orthonormal series in $L_2(\mu)$ into a similar theory for series in noncommutative L_p-spaces $L_p(\mathcal{M}, \varphi)$ or even symmetric spaces $E(\mathcal{M}, \varphi)$ constructed over a noncommutative measure space $(\mathcal{M}, \varphi)$, a von Neumann algebra $\mathcal{M}$ of operators acting on a Hilbert space H together with a faithful normal state φ on this algebra.

In Chap. 2 we present a new and modern understanding of the classical theory on pointwise convergence of orthonormal series in the Hilbert spaces $L_2(\mu)$, and show that large parts of the classical theory transfer to a theory on pointwise convergence of unconditionally convergent series in spaces $L_p(\mu, X)$ of μ-integrable functions with values in Banach spaces X, or more generally Banach function spaces $E(\mu, X)$ of X-valued μ-integrable functions. Here our tools are strongly based on Grothendieck's metric theory of tensor products and in particular on his théorème fondamental. In Chap. 3 this force turns out to be even strong enough to extend our scheme to the setting of symmetric spaces $E(\mathcal{M}, \varphi)$ of operators and Haagerup L_p-spaces $L_p(\mathcal{M}, \varphi)$. In comparison with the old classical commutative setting the new noncommutation setting highlights new phenomena, and our theory as a whole unifies, completes and extends various results, both in the commutative and in the noncommutative world.

Oldenburg, Germany   *Andreas Defant*

Contents

Chapter 1
Introduction

In the theory of orthogonal series the most important coefficient test for almost
everywhere convergence of orthonormal series is the fundamental theorem of Menchoff and Rademacher which was obtained independently in [60] and [81]. It states
that whenever a sequence (α_k) of coefficients satisfies the "test" $\sum_k |\alpha_k \log k|^2 < \infty$,
then for every orthonormal series $\sum_k \alpha_k x_k$ in $L_2(\mu)$ we have that

$$\sum_k \alpha_k x_k \text{ converges } \mu\text{-almost everywhere.} \tag{1.1}$$

The aim of this research is to develop a systematic scheme which allows us
to transform important parts of the now classical theory of almost everywhere
summation of orthonormal series in $L_2(\mu)$ into a similar theory for series in noncommutative L_p-spaces $L_p(\mathcal{M}, \varphi)$ or even symmetric spaces $E(\mathcal{M}, \varphi)$ buildup over
a noncommutative measure space $(\mathcal{M}, \varphi)$, a von Neumann algebra $\mathcal{M}$ of operators
acting on a Hilbert space H together with a faithful normal state φ on this algebra.

The theory of noncommutative L_p-spaces – the analogs of ordinary Legesgue
spaces $L_p(\mu)$ with a noncommutative von Neumann algebra playing the role of
$L_\infty(\mu)$ – for semifinite von Neumann algebras was laid out in the early 1950s
by Dixmier [10], Kunze [49], Segal [84], and Nelson [66]. Later this theory was
extended in various directions. In a series of papers P.G. Doods, T.K. Doods and
de Pagter [11, 12, 13, 14], Ovchinnikov [69, 70], Chilin and Sukochev [88], Kalton
and Sukochev [45], Sukochev [85, 86, 87, 88] and Xu [96] developed a method for
constructing symmetric Banach spaces of measurable operators, and Haagerup in
[24] presented a theory of L_p-spaces associated with not necessarily semifinite von
Neumannn algebras. See [80] for a beautiful survey on the whole subject.

In the 1980s the school of Łódź started a systematic study, mainly due to Hensz
and Jajte, of Menchoff-Rademacher type theorems within the Gelfand-Naimark
construction $L_2(\mathcal{M}, \varphi)$, the natural Hilbert space constructed over a von Neumann
algebra $\mathcal{M}$ and a faithful normal state φ on $\mathcal{M}$, and most of their results were
later collected in the two Springer Lecture Notes [37, 38]. Based on new techniques
mainly due to Junge, Pisier, and Xu (see e.g. [39, 42, 79]) some of these results were

A. Defant, *Classical Summation in Commutative and Noncommutative Lp-Spaces,*
Lecture Notes in Mathematics 2021, DOI 10.1007/978-3-642-20438-8_1,
© Springer-Verlag Berlin Heidelberg 2011

generalized in [7], and we plan to use similar tools to complement and extend this work in various directions.

In a *first step* we give a new and modern understanding of the classical theory on pointwise convergence of general orthonormal series in the Hilbert spaces $L_2(\mu)$. Then in a *second step* we show that large parts of the classical theory transfer to a theory on pointwise convergence of unconditionally convergent series in spaces $L_p(\mu, X)$ of μ-integrable functions with values in Banach spaces X, or more generally Banach function spaces $E(\mu, X)$ of X-valued μ-integrable functions. Here our tools are strongly based on Grothendieck's metric theory of tensor products and in particular on his *théorème fondamental* – this force in a *third step* is even strong enough to extend our scheme to the setting of symmetric spaces $E(\mathscr{M}, \varphi)$ of operators and Haagerup L_p-spaces $L_p(\mathscr{M}, \varphi)$. It turns out that in comparison with the old classical commutative setting the new noncommutative setting highlights new phenomena, and our theory as a whole unifies, completes and extends the original theory, both in the commutative and in the noncommutative setting:

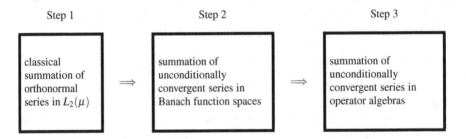

1.1   Step 1

Let us start by describing more carefully some of the prototypical results of the classical theory of almost everywhere summation of general orthogonal series which form the historical background of these notes; as standard references we use the monographs [1] and [47]. See again the Menchoff-Rademacher theorem from (1.1) which is the most fundamental one. As quite often happens in Fourier analysis, this fact comes with an at least formally stronger result on maximal functions. Kantorovitch in [46] proved that the maximal function (of the sequence of partial sums) of each orthonormal series $\sum_k \alpha_k x_k$ in $L_2(\mu)$ satisfying $\sum_k |\alpha_k \log k|^2 < \infty$, is square-integrable,

$$\sup_j \left| \sum_{k=0}^{j} \alpha_k x_k \right| \in L_2(\mu), \tag{1.2}$$

or in other terms: For each choice of finitely many orthonormal functions $x_0, \ldots, x_n \in L_2(\mu)$ and each choice of finitely many scalars $\alpha_0, \ldots \alpha_n$ we have

$$\left\| \sup_j \left| \sum_{k=0}^{j} \alpha_k x_k \right| \right\|_2 \leq C \left(\sum_{k=0}^{n} |\alpha_k \log k|^2 \right)^{1/2}, \tag{1.3}$$

where C is an absolute constant. This inequality is also known as the Kantorovitch-Menchoff-Rademacher inequality.

Results of this type can and should be seen in the context of limit theorems for random variables. To illustrate this by an example, we mention that by a simple lemma usually attributed to Kronecker, the Menchoff-Rademacher theorem (1.1) implies the following strong law of large numbers: Given a sequence of uncorrelated random variables X_k on a probability space (Ω, P) we have that

$$\left(\frac{1}{j+1} \sum_{k=0}^{j} X_k - E(X_k) \right)_j \text{ converges } P\text{-almost everywhere}, \tag{1.4}$$

provided

$$\sum_k \frac{\log^2 k}{k^2} \operatorname{Var}(X_k) < \infty.$$

There is a long list of analogs of the Menchoff-Rademacher theorem for various classical summation methods, as e.g. Cesàro, Riesz, or Abel summation. Recall that a summation method formally is a scalar matrix $S = (s_{jk})$ with positive entries such that for each convergent series $s = \sum_n x_n$ of scalars we have

$$s = \lim_j \sum_k s_{jk} \sum_{\ell=0}^{k} x_\ell. \tag{1.5}$$

Let us explain what we mean by *a coefficient test for almost everywhere convergence of orthonormal series* with respect to a summation method $S = (s_{kj})$ and an increasing and unbounded sequence (ω_k) of positive scalars. We mean a theorem which assures that

$$\sum_k \alpha_k x_k = \lim_j \sum_k s_{jk} \sum_{\ell=0}^{k} \alpha_\ell x_\ell \quad \mu\text{-a.e.} \tag{1.6}$$

whenever $\sum_k \alpha_k x_k$ is an orthonormal series in some $L_2(\mu)$ with coefficients (α_k) satisfying the *test* $\sum_k |\alpha_k \omega_k|^2 < \infty$; in the literature such a sequence ω is then often called a Weyl sequence.

For example, the coefficient test $\sum_k |\alpha_k \log \log k|^2 < \infty$ assures that all orthonormal series $\sum_k \alpha_k x_k$ are almost everywhere Cesàro-summable, i.e.

$$\sum_k \alpha_k x_k = \frac{1}{j+1} \sum_{k=0}^{j} \sum_{\ell=0}^{k} \alpha_\ell x_\ell \quad \mu\text{-a.e.}, \tag{1.7}$$

and moreover the maximal function of the arithmetic means satisfies

$$\sup_j \left| \frac{1}{j+1} \sum_{k=0}^{j} \sum_{\ell=0}^{k} \alpha_\ell x_\ell \right| \in L_2(\mu); \tag{1.8}$$

this is a famous result of Kaczmarz [43] and Menchoff [61, 62].

Our main new contribution on step 1 is to analyze the proofs of some classical coefficient tests on Cesàro, Riesz or Abel summation in order to show that they all include a maximal theorem of the form given in (1.2) or (1.8): For each orthonormal sequence (x_k) in $L_2(\mu)$ and each $\alpha \in \ell_2$ the maximal function of the linear means of partial sums of the series $\sum_k \alpha_k x_k$ is square integrable,

$$\sup_j \left| \sum_k s_{jk} \sum_{\ell=0}^{k} \frac{\alpha_\ell}{\omega_\ell} x_\ell \right| \in L_2(\mu). \tag{1.9}$$

Moreover, we will see as a consequence of the well-known dilation theorem of Nagy that such a maximal theorem then automatically holds for a much larger class of unconditionally summable sequences (x_k) than the orthonormal ones, namely for all so called weakly 2-summable sequences (x_k) in $L_2(\mu)$ (i.e. sequences such that for all $\alpha \in \ell_2$ the series $\sum_k \alpha_k x_k$ converges in $L_2(\mu)$).

1.2   Step 2

In the 1970s it became clear that several of these classical coefficient tests have remarkable extensions to the summation of unconditionally convergent series $\sum_k x_k$ in Banach spaces $L_p(\mu)$, μ some measure and $1 \leq p < \infty$. The following counterpart of the Menchoff-Rademacher theorem was proved by Bennett [2] and independently Maurey-Nahoum [59]: Take $\sum_k x_k$ an unconditionally convergent series (i.e. each rearrangement of the series converges as well) in $L_p(\mu)$, $1 \leq p < \infty$. Then the maximal function of its partial sums is p-integrable up to a log-term,

$$\sup_j \left| \sum_{k=0}^{j} \frac{x_k}{\log k} \right| \in L_p(\mu), \tag{1.10}$$

and as a consequence

$$\sum_k \frac{x_k}{\log k} \text{ converges } \mu\text{-almost everywhere}; \tag{1.11}$$

throughout these notes read $\log x$ *(base 2) for* $\max\{1, \log x\}$. The crucial case $p = 1$ had already been shown slightly earlier by Maurey in [58] – it had been conjectured by Kwapień and Pełczyński in [50] who proved a weaker form. Since each orthonormal series $\sum_k \alpha_k x_k$ in $L_2(\mu)$ is of course unconditionally convergent, the case $p = 2$ of this theorem then still contains its "godfather", the Menchoff-Rademacher theorem, as a special case.

As explained above a classical coefficient test like in (1.6) usually is a consequence of a maximal theorem as in (1.9). In Chap. 2 we are going to show that a summation method S satisfies a maximal theorem of this type if and only if the infinite matrix $A = (a_{jk})$ given by

$$a_{jk} := \frac{1}{\omega_k} \sum_{\ell=k}^{\infty} s_{j\ell}$$

and viewed as a (bounded and linear) operator from ℓ_∞ into ℓ_1 factorizes through a Hilbert space (such operators are called hilbertian). On the other hand Grothendieck's théorème fondamental (almost in its original form from [21]) states that each operator $A : \ell_\infty \to \ell_1$ factorizes through a Hilbert space if and only if it allows the (a priori much stronger) factorization

$$
\begin{array}{ccc}
\ell_\infty & \xrightarrow{\ A\ } & \ell_1 \\[2pt]
\downarrow{\scriptstyle u} & & \uparrow{\scriptstyle v} \\[2pt]
L_\infty(\mu) & \xhookrightarrow[\ j\ ]{} & L_1(\mu),
\end{array}
$$

where u, v are operators and μ a probability measure. Now, if A factorizes this way – Grothendieck called such operators A integral – then it is just a small step to prove that for each unconditionally convergent series $\sum_k x_k$ in any $L_1(\mu)$ we even get

$$\sup_j \left| \sum_k s_{jk} \sum_{\ell=0}^{k} \frac{x_\ell}{\omega_k} \right| \in L_1(\mu). \tag{1.12}$$

This is the L_1-counterpart of (1.9). The fact that an operator $u : \ell_\infty \to \ell_1$ is hilbertian if and only if it is integral (Grothendieck's théorème fondamental), explains from an abstract point of view why the classical theory on pointwise summation of general orthonormal series has an in a sense equivalent L_p-theory.

The aim of the second chapter is to introduce our setting of so called maximizing matrices (see the definition from Sect. 2.1.3). This allows us to transform basically every classical coefficient test for almost everywhere convergence of orthonormal series $\sum_k \alpha_k x_k$ in $L_2(\mu)$ to a theorem on almost everywhere summation for unconditionally convergent series $\sum_k x_k$ in vector-valued Banach function spaces $E(\mu, X)$ (all μ-measurable functions x with values in a Banach space X such that $\|x(\cdot)\|_X \in E(\mu)$).

More precisely, assume that we have a coefficient test for a summation method $S = (s_{kj})$ and a Weyl sequence ω which additionally fulfills the maximal theorem from (1.9). Then a considerably stronger result holds: For every unconditionally convergent series $\sum_k x_k$ in an arbitrary vector-valued Banach function space $E(\mu, X)$ we have

$$\sup_j \left\| \sum_k s_{jk} \sum_{\ell=0}^{k} \frac{x_\ell(\cdot)}{\omega_\ell} \right\|_X \in E(\mu), \tag{1.13}$$

and as a consequence

$$\sum_k \frac{x_k}{\omega_k} = \lim_j \sum_k s_{jk} \sum_{\ell=0}^{k} \frac{x_\ell}{\omega_\ell} \quad \mu\text{-}a\text{-}e.. \tag{1.14}$$

Again, since each orthonormal series $\sum_k \alpha_k x_k$ in L_2 is unconditionally summable, each such $E(\mu, X)$-extension of a coefficient test still contains its original source as a special case.

For some particular summation methods S, as e.g. Cesàro or Riesz summation, such results for $L_p(\mu)$-spaces have been known for almost forty years, and the technical heart of the machinery which makes this possible is the theory of p-summing and p-factorable operators – mainly through fundamental contributions of Grothendieck, Kwapień and Pietsch from the fifties and sixties. The initial ideas were born in the papers [50] and [83] of Kwapień-Pełczyński and Schwartz, and the highlights came in the seventies with articles of Bennett [2, 3], Maurey [58], Maurey-Nahoum [59], and Orno [68]. We would like to emphasize that much of the content of the second chapter was deeply inspired by Bennett's two masterpieces [2, 3]. In [3] Bennett found an elementary approach to several limit theorems in L_p-spaces proving them independently of the classical theory of orthonormal series. This as a consequence lead him to the following provocative sentence in the introduction of [3]: " ... the theory of orthonormal series has little to do with orthonormal series ..." .

We have a less radical point of view. Our approach is a mixture of all the methods mentioned so far. It uses the original theory on pointwise convergence of general orthogonal series as a starting point, and then inspired by [3] relates it with modern operator theory in Banach spaces in order to deduce new results in more general, in particular noncommutative, settings.

1.3   Step 3

Let us come back to the noncommutative situation. The third chapter which originally motivated all of this research, shows that our theory of maximizing matrices is powerful enough to convert major parts of the classical theory on pointwise summation of orthonormal series into a theory on summation of unconditionally convergent series in noncommutative L_p-spaces or symmetric spaces of operators.

Fix some von Neumann algebra $\mathcal{M}$ of operators acting on a Hilbert space H, and a normal and faithful state φ on $\mathcal{M}$; the pair

$$(\mathcal{M}, \varphi)$$

will be called a noncommutative probability space. Every usual probability space (Ω, Σ, μ) defines such a pair $(\mathscr{M}, \varphi)$ (through $\mathscr{M} = L_\infty(\mu)$ and $\varphi(x) = \int x d\mu$, $x \in L_\infty(\mu)$).

The remainder of this introduction is divided into three parts. Mainly guided by the classical commutative theory (part of which is presented in Chap. 2), we try to convince our reader that a systematic study of noncommutative coefficient tests in the following three different settings is natural:

- Symmetric spaces $E(\mathscr{M}, \varphi)$ whenever φ is tracial,
- Haagerup's spaces $L_p(\mathscr{M}, \varphi)$,
- And the algebra $(\mathscr{M}, \varphi)$ itself.

Within these three settings in reverse order we now collect some natural questions, and anticipate some of the answers we intend to give.

1.3.1 Coefficient Tests in the Algebra Itself

An important step in the famous GNS-construction is that the definition

$$(x|y) := \varphi(y^*x), \quad x, y \in \mathscr{M} \tag{1.15}$$

leads to a scalar product on $\mathscr{M}$, and hence

$$\|x\|_2 = \varphi(x^*x)^{1/2}, \quad x \in \mathscr{M}$$

gives a hilbertian norm on $\mathscr{M}$. We write $\|x\|_\infty$ for the norm in $\mathscr{M}$ in order to distinguish it from this norm. With the noncommutative theory in mind it is then obvious what is meant by a φ-orthonormal series $\sum_k \alpha_k x_k$ in $\mathscr{M}$: the sequence of all α_k's belongs to ℓ_2 and the x_k form an orthonormal system in the prehilbertian space $(\mathscr{M}, \|\cdot\|_2)$.

Several interesting questions appear:

(I) Is there any analog of the Menchoff-Rademacher theorem from (1.1) for such φ-orthonormal series? In which sense do φ-orthonormal series $\sum_k \alpha_k x_k$ converge almost everywhere?

(II) Is there any analog of the Menchoff-Rademacher-Kantorovitch maximal inequality from (1.3) in this setting of operator algebras?

(III) What if ordinary summation like in the Menchoff-Rademacher theorem (1.1) is replaced by some other classical summation methods? For example, what about an analog of the Kaczmarcz-Menchoff result for Cesàro summation from (1.7) and (1.8)?

(IV) If any such results exist, do they, like in Maurey's theorem from (1.10), in any sense transfer to the predual of $\mathscr{M}$, the noncommuative analog of $L_1(\mu)$?

As mentioned before, coefficient tests from the commutative world typically are consequences of maximal inequalities, hence let us first discuss question II. A straightforward transformation to the noncommutative setting of the maximal function $\sup_k |f_k|$ of a sequence of functions f_k in $L_2(\mu)$ is not possible. Even for two positive operators x_1 and x_2 on a Hilbert space H there may be no operator x such that taking scalar products we have $(x\xi|\xi) = \sup_{k=1,2}(x_k\xi|\xi)$ for all $\xi \in H$. But although we cannot offer something like a maximal operator $\sup_k x_k$ itself, we have a reasonable substitute in the hilbertian norm $\| \sup_k x_k \|_2$ of such an object.

In the commutative setting we obviously have that for each choice of finitely many functions $f_0, \ldots, f_k \in L_2(\mu)$

$$\Big\| \sup_k |f_k| \Big\|_2 = \inf \sup_k \|z_k\|_\infty \|c\|_2 \,,$$

the infimum taken over all possible decompositions $f_k = z_k c$ by bounded functions z_k and a square integrable function c. This formula is the model for the following definition: Consider an $(n+1)$-tuple $(x_0, \ldots, x_n)$ of operators in $\mathcal{M}$, and put

$$\||(x_k)\|| := \inf \sup_k \|z_k\|_\infty \|c\|_2 \,, \tag{1.16}$$

the infimum now taken over all $z_k, c \in \mathcal{M}$ such that $x_k = z_k c$ for all k. This leads to a norm on the vector space of all these $(n+1)$-tuples, and as a consequence it allows us to formulate and to prove as a perfect analog of (1.3) the following Kantorovitch-Menchoff-Rademacher inequality in noncommutative probability spaces $(\mathcal{M}, \varphi)$: For each choice of finitely many φ-orthonormal operators $x_0, \ldots, x_n \in \mathcal{M}$ and each choice of finitely many scalars $\alpha_0, \ldots \alpha_n$ we have

$$\Big\|\Big(\sum_{k=0}^{j} \alpha_k x_k \Big)_{0 \le j \le n}\Big\| \le C\Big(\sum_{k=0}^{n} |\alpha_k \log k|^2 \Big)^{1/2} \,, \tag{1.17}$$

where C is an absolute constant.

What about question I – in which sense do φ-orthonormal series $\sum_k \alpha_k x_k$ converge almost everywhere? For noncommutative probability spaces $(\mathcal{M}, \varphi)$ there is a natural substitute for almost everywhere convergence of sequences in $\mathcal{M}$. The model for its definition comes from Egoroff's theorem. By this theorem a sequence (f_n) of μ-measurable functions converges to 0 μ-almost everywhere whenever

$$f_n \to 0 \ \mu\text{-almost uniformly}, \tag{1.18}$$

i.e. for every $\varepsilon > 0$ there is a measurable set A such that $\mu(\Omega \setminus A) < \varepsilon$ and $(f_n\chi_A)$ converges to 0 uniformly, or in other terms, $(f_n\chi_A)$ converges to 0 in $L_\infty(\mu)$. Since the indicator functions χ_A in $L_\infty(\mu)$ are precisely the projections p from the von Neumann algebra $L_\infty(\mu)$, the following well-known and widely used definition is a perfect analog of almost everywhere convergence in noncommutative probability spaces $(\mathcal{M}, \varphi)$: A sequence (x_n) in $\mathcal{M}$ converges to 0 φ-almost uniformly (is a φ-almost uniform Cauchy sequence) whenever for every $\varepsilon > 0$ there is a projection p in $\mathcal{M}$ such that $\varphi(1-p) < \varepsilon$ and $(x_n p)$ converges to 0 in $\mathcal{M}$ (is a Cauchy sequence in $\mathcal{M}$).

To the best of our knowledge the only known result on Menchoff-Rademacher type theorems within the setting of φ-orthonormal series in the prehilbertian space $(\mathcal{M}, \|\cdot\|_2)$ is the following noncommutative law of large numbers proved by Jajte [37, Sect. 4.4.1], an analog of (1.4): Let (x_k) be a φ-orthogonal sequence in $\mathcal{M}$ such that $\sum_k \frac{\log^2 k}{k} \|x_k\|_2^2 < \infty$. Then

$$\lim_j \frac{1}{j+1} \sum_{k=0}^{j} x_k = 0 \quad \varphi\text{-almost uniformly.} \tag{1.19}$$

We will get much more – e.g. in Sect. 3.2 we show as an analog of (1.1) that for each φ-orthonormal series $\sum_k \alpha_k x_k$ in $\mathcal{M}$ such that $\sum_k |\alpha_k \log k|^2 < \infty$ the sequence

$$\left(\sum_{k=0}^{j} \alpha_k x_k\right)_j \quad \text{is } \varphi\text{-almost uniform Cauchy,} \tag{1.20}$$

and if in the prehilbertian space $(\mathcal{M}, \|\cdot\|_2)$ the Cauchy sequence $\sum_k \alpha_k x_k$ converges to some $x \in \mathcal{M}$ (and satisfies an additional technical condition), then we even have that

$$\sum_{k=0}^{\infty} \alpha_k x_k = x \quad \varphi\text{-almost uniformly.} \tag{1.21}$$

In fact, in Sects. 3.2.7 and 3.2.9 we will give reasonable answers to all of the questions (I)–(IV) given above. Why is (1.20) just a result on φ-almost uniform Cauchy sequences and not φ-almost uniform convergent sequences? The commutative case $\mathcal{M} = L_\infty(\mu)$ immediately shows that we in general cannot expect that $\sum_k \alpha_k x_k$ converges μ-almost everywhere to some element from $L_\infty(\mu)$ itself. This partly motivates the following topic.

1.3.2   Tests in the Hilbert Spaces $L_2(\mathcal{M}, \varphi)$ and More Generally Haagerup L_p's

The results in (1.20) and (1.17) only deal with φ-orthonormal series in the prehilbertian space $(\mathcal{M}, \|\cdot\|_2)$. What about summation of orthonormal series in the GNS-construction $L_2(\mathcal{M}, \varphi)$ (a Hilbert space) itself?

Given a noncommutative probability space $(\mathcal{M}, \varphi)$, the completion of $(\mathcal{M}, \|\cdot\|_2)$ gives the Hilbert space $L_2(\mathcal{M}, \varphi)$, the GNS-representation space for $\mathcal{M}$ with respect to φ. Recall that $\mathcal{M}$ by left multiplication acts on this Hilbert space $L_2(\mathcal{M}, \varphi)$

with a cyclic and separating vector, denoted by $D^{1/2} \in L_2(\mathcal{M}, \varphi)$, such that $\varphi(x) = (xD^{1/2}|D^{1/2})$ for each $x \in \mathcal{M}$. As usual, we identify $\mathcal{M}$ with the subspace $\mathcal{M}D^{1/2}$ of $L_2(\mathcal{M}, \varphi)$.

What does an analog of the Menchoff-Rademacher theorem or other classical coefficient tests in the GNS-representation space $L_2(\mathcal{M}, \varphi)$ look like? The first attempt to study the Menchoff-Rademacher theorem in this noncommutative setting was made by Hensz and Jajte; together with their coauthors they published a series of papers on this topic, and as already mentioned above most of these results were collected in the two Springer lecture notes [37] and [38].

Since $L_2(\mathcal{M}, \varphi)$ is defined by completion, it is now a bit more difficult to give a reasonable definition of almost everywhere convergence of a sequence in $L_2(\mathcal{M}, \varphi)$. Again the model for this definition comes from Egoroff's theorem: A sequence (x_n) in $L_2(\mathcal{M}, \varphi)$ converges to 0 φ-almost surely whenever for every $\varepsilon > 0$ there is a projection p in $\mathcal{M}$ with $\varphi(1 - p) < \varepsilon$ and a matrix (x_{nk}) with entries in $\mathcal{M}$ such that

$$\sum_{k=0}^{\infty} x_{nk}D^{1/2} = x_n \text{ in } L_2(\mathcal{M}, \varphi) \text{ and } \left\|\sum_{k=0}^{\infty} x_{nk}p\right\|_{\infty} \to 0.$$

Let us define the analog of (1.16) for $(n+1)$-tuples $(x_0, \ldots, x_n)$ in $L_2(\mathcal{M}, \varphi)$:

$$|||(x_k)||| := \inf \sup_k \|z_k\|_{\infty}\|c\|_2 \qquad (1.22)$$

the infimum taken over all uniform factorizations $x_k = z_k c$ with $z_k \in \mathcal{M}$ and $c \in L_2(\mathcal{M}, \varphi)$. Again we get a norm on all these $(n+1)$-tuples, and the Menchoff-Rademacher-Kantorovitch inequality from (1.3) reads as follows: There is a constant $C > 0$ such that for each choice of finitely many orthonormal operators $x_0, \ldots, x_n \in L_2(\mathcal{M}, \varphi)$ and scalars $\alpha_0, \ldots, \alpha_n$ we have

$$\left|\left|\left|\left(\sum_{k=0}^{j} \alpha_k x_k\right)_{0 \le j \le n}\right|\right|\right| \le C\left(\sum_{k=0}^{n} |\alpha_k \log k|^2\right)^{1/2}. \qquad (1.23)$$

From this noncommutaive maximal inequality we then obtain one of the major results from the study in [38], the following analog of the Menchoff-Rademacher theorem (1.1): if the coefficients of an orthonormal series $\sum_k \alpha_k x_k$ in $L_2(\mathcal{M}, \varphi)$ satisfy $\left(\sum_k |\alpha_k \log k|^2\right)^{1/2} < \infty$, then

$$\sum_k \alpha_k x_k \text{ converges } \varphi - \text{almost surely}.$$

But in Sect. 3.2 of Chap. 3 we are able to do much more. We are going to consider Haagerup L_p-spaces $L_p(\mathcal{M}, \varphi)$ instead of $L_2(\mathcal{M}, \varphi)$ only, and we will consider various summation methods instead of only ordinary summation. Most of our results are formulated for unconditionally convergent series instead of orthonormal series, so for a much richer class of series. Most of our noncommutative coefficient tests follow from related noncommutative maximal theorems, and a comparison with the commutative situation studied in Chap. 2 shows new phenomena. Concrete examples show that we have to distinguish carefully between the cases $p \leq 2$ and $p > 2$. This leads to different types of maximal theorems and different types of theorems on almost everywhere convergence.

1.3.3   Tests in Symmetric Spaces $E(\mathcal{M}, \tau)$, τ a Trace

For a state φ on a von Neumann algebra $\mathcal{M}$ the Hilbert space $L_2(\mathcal{M}, \varphi)$ is defined to be the completion of $(\mathcal{M}, \| \cdot \|_2)$ – so one cannot really touch the objects in $L_2(\mathcal{M}, \varphi)$. The setting gets far richer if we assume φ to be a trace instead of only a state. Let us now fix a noncommutative integration space

$$(\mathcal{M}, \tau),$$

$\mathcal{M}$ a semifinite von Neumann algebra and τ a normal, faithful and semifinite trace τ. According to the work of Dixmier [10], Nelson [66], Segal [84], and others the Banach spaces $L_p(\mathcal{M}, \tau)$ can then be realized as subspaces of the space $L_0(\mathcal{M}, \tau)$. This space collects all τ-measurable operators on H affiliated to $\mathcal{M}$, operators which in many respects behave very much like the operators in the algebra itself. The noncommutative L_p-spaces $L_p(\mathcal{M}, \tau)$ then consist of all $x \in L_0(\mathcal{M}, \tau)$ for which $\|x\|_p := \tau(|x|^p)^{1/p} < \infty$. By definition $\mathcal{M}$ equals $L_\infty(\mathcal{M}, \tau)$ and $L_1(M, \tau)$ turns out to be the predual $\mathcal{M}_*$ of $\mathcal{M}$.

In $L_0(\mathcal{M}, \tau)$ τ-almost uniform convergence is defined as follows: (x_n) converges in $L_0(\mathcal{M}, \tau)$ τ-almost uniformly to $x \in L_0(\mathcal{M}, \tau)$ if for each $\varepsilon > 0$ there is a projection p in $\mathcal{M}$ such that $\tau(1 - p) < \varepsilon$, $(x_n - x)p \in \mathcal{M}$ for all n and $\|(x_n - x)p\|_\infty \to 0$ as $n \to \infty$. Moreover, $x_n \to x$ bilaterally τ-almost uniformly provided for each ε there is p such $p(x_n - x)p \in \mathcal{M}$ and $\|p(x_n - x)p\|_\infty \to 0$.

The first study of noncommutative L_p-variants, $p \neq 2$ of the Menchoff-Rademacher theorem was given in [7], a paper which contains the origin and very heart of the work described here. The main result from [7] states that for each unconditionally convergent series in $L_p(\mathcal{M}, \tau)$ the series

$$\sum_k \frac{x_k}{\log k} \text{ converges } \begin{cases} \text{bilaterally } \tau\text{-almost uniformly} & p \leq 2 \\ \tau\text{-almost uniformly} & p \geq 2. \end{cases} \qquad (1.24)$$

Clearly, this result is a noncommutative extension of the original Menchoff-Rademacher theorem from (1.1) and its extensions by Bennett, Maurey and Maurey-Nahoum given in (1.11). What about the maximal theorem behind it? Similar to (1.16) and (1.22) we define

$$L_p(\mathcal{M}, \tau)[\ell_\infty] \qquad (1.25)$$

as the space of all sequences (x_n) in $L_p(\mathcal{M}, \tau)$ such that there exist $c, d \in L_{2p}(\mathcal{M}, \tau)$ and a bounded sequence (z_n) in $\mathcal{M}$ for which $x_n = cz_n d$ for all n, and the symbol $L_p(\mathcal{M}, \tau)[\ell_\infty^c]$ (here the additional letter c stands for column) denotes all sequences (x_n) which allow a factorization $x_n = z_n d$ with $d \in L_p(\mathcal{M}, \tau)$ and (z_n) uniformly bounded (see Sect. 3.1.2). With this notion the maximal theorem from (1.10) extends as follows:

$$\left(\sum_{k=0}^{j} \frac{x_k}{\log k} \right)_j \in \begin{cases} L_p(\mathcal{M}, \tau)[\ell_\infty] & p \geq 2 \\ L_p(\mathcal{M}, \tau)[\ell_\infty^c] & p \leq 2. \end{cases}$$

We improve this result in several directions within the setting of symmetric spaces $E(\mathcal{M}, \tau)$. Recall the definition of a symmetric space of operators: Given a symmetric Banach function space $(E, \|\cdot\|_E)$ on the interval $[0, \tau(1))$ the symmetric space of operators with respect to the integration space $(\mathcal{M}, \tau)$ and E is defined to be

$$E(\mathcal{M}, \tau) := \left\{ x \in L_0(\mathcal{M}, \tau) \,\middle|\, \mu(x) \in E \right\};$$

here $\mu(x)$ as usual denotes the decreasing rearrangement of x. Looking at (1.25) (or the definitions given in (1.16) and (1.22)) it is now fairly straightforward to find the right definition of $E(\mathcal{M}, \tau)[\ell_\infty]$ (see again Sect. 3.1.2).

In the setting of symmetric spaces of operators we get almost perfect analogs of the related commutative results; we e.g. show that whenever a coefficient test for S and ω is given which *additionally* fulfills (1.9), then for every unconditionally convergent series $\sum_k x_k$ in $E(\mathcal{M}, \tau)$ we have that

$$\left(\sum_k s_{jk} \sum_{\ell=0}^{k} \frac{x_\ell}{\omega_\ell} \right)_j \in E(\mathcal{M}, \tau)[\ell_\infty], \tag{1.26}$$

and as a consequence

$$\sum_k \frac{x_k}{\omega_k} = \lim_j \sum_k s_{jk} \sum_{\ell=0}^{k} \frac{x_\ell}{\omega_\ell} \quad \text{bilaterally } \tau\text{-almost uniformly}. \tag{1.27}$$

Moreover, under certain geometric assumption on the function space E (convexity or concavity conditions) the result on bilateral τ-almost uniform convergence as well as the maximal theorem improve. Based on the results from the second chapter these results apply to Cesàro, Riesz, and Abel summation.

Let us finally remark that the crucial definitions in (1.16), (1.22), and (1.25) are motivated by the work on vector-valued noncommutative L_p-spaces due to to Pisier from [79], and Junge's article [39] in which he proves his celebrated noncommutative version of Doob's inequality. In view of the unconditionality of noncommutative martingale difference sequences (see e.g. [80, Corollary 7.3]) our method implies the noncommutative Doob inequality only up to a logarithmic factor – but on the other hand our techniques and our results apply to some far more general situations.

1.4   Preliminaries

We shall use standard notation and notions from Banach space theory as presented e.g. in [53, 54, 92, 94]. The unit ball of a Banach space X (over the scalar field $\mathbb{K} = \mathbb{R}$ or $\mathbb{C}$) is denoted by B_X, $\mathscr{L}(X,Y)$ stands for the Banach space of all (bounded and linear) operators between two Banach spaces X and Y, and X' for the dual of X. Clearly, in combination with von Neumann algebras we only consider complex Banach spaces.

A lot of information will be given locally (see also the subject and symbol index). For the definition of what we call Banach function spaces $E(\mu, X)$ see the beginning of Sect. 2.1.2; in particular, for the definition of ℓ_p and ℓ_p^n (in these notes the latter space is $n + 1$-dimensional, the vector space $\mathbb{K}^{n+1}$ together with the norm $\|x\|_p = (\sum_{j=0}^n |x_j|^p)^{1/p}$, $1 \le p < \infty$ and $\|x\|_\infty = \sup_{0 \le j \le n} |x_j|$, $p = \infty$). As usual, we write p' for the conjugate of $1 \le p \le \infty$ defined by $1/p + 1/p' = 1$. The symbol S_p stands for the Schatten p-class. See Sect. 2.1.3 for the definition of unconditionally summable and of weakly p-summable sequences in a Banach space X together with the corresponding Banach spaces $(\ell_1^{\mathrm{unc}}(X), w_1)$ and $(\ell_p^w(X), w_p)$. For the definition of convex and concave Banach lattices see Sect. 3.1.1 (and also (2.68)), and for the definition of Banach spaces with cotype we refer to (2.71).

Most of the information we need on the theory of almost everywhere convergence of orthogonal series is in fact contained in Sect. 2.2; our standard references are [1, 47, 97]. The needed basics of the theory of Banach operator ideals which is intimately related with Grothendieck's metric theory of tensor products are repeated in Sect. 2.1.6; see the monographs [6, 9, 76, 77]. In the tracial case a brief introduction to noncommutative L_p-spaces and more generally symmetric spaces of measurable operators is given in Sect. 3.1.1, and in the nontracial case on Haagerup L_p-spaces in Sect. 3.2.7; the references we use come mostly from [37, 38, 65, 80, 89, 90].

For $x \in \mathbb{R}_{\ge 0}$, the logarithm $\log x$ is always taken with respect to the base 2, and we repeat that throughout these notes $\log x$ is understood as $\max\{1, \log x\}$.

Acknowledgment

It is a great pleasure to thank my friend and colleague Marius Junge who was always willing and able to discuss my problems. Without his help these notes would be very different. I am also grateful to Pablo Sevilla Peris for his great support.

Chapter 2
Commutative Theory

2.1 Maximizing Matrices

We invent a class of infinite matrices $A = (a_{jk})_{j,k=0}^{\infty}$ called (p,q)-maximizing; its definition (see Definition 1 in Sect. 2.1.3) is motivated by a number of classical maximal inequalities intimately related with almost sure summation of orthogonal series with respect to Cesàro, Riesz, and Abel summation. The main examples (given in the next section) are matrix products $A = S\Sigma$ and their "diagonal perturbations" $S\Sigma D_{1/\omega}$, where S is a summation process (see (1.5)), $\Sigma = (\sigma_{jk})$ the so-called sum matrix defined by

$$\sigma_{jk} = \begin{cases} 1 & k \leq j \\ 0 & k > j, \end{cases} \tag{2.1}$$

and $D_{1/\omega}$ the diagonal matrix with respect to a Weyl sequence ω. Recall that an increasing and unbounded sequence (ω_k) of positive scalars is said to be a Weyl sequence with respect to a summation method $S = (s_{jk})$ whenever for each orthonormal series in $L_2(\mu)$ we have that

$$\sum_k \alpha_k x_k = \lim_j \sum_k s_{jk} \sum_{\ell=0}^{k} \alpha_\ell x_\ell \quad \mu\text{-a. e.} \tag{2.2}$$

provided $\sum_k |\alpha_k \omega_k|^2 < \infty$; as already explained in (1.6) we call a theorem of this type a coefficient test.

Based on Nagy's dilation lemma we in Theorem 1 characterize $(2,2)$-maximizing matrices in terms of orthonormal series in $L_2(\mu)$, a result which later in Sect. 2.2 will turn out to be crucial in order to derive non trivial examples of maximizing matrices from classical coefficient tests. Theorem 2 shows that for $q < p$ every matrix product $S\Sigma$ is (p,q)-maximizing, whereas for $q \geq p$ an

A. Defant, *Classical Summation in Commutative and Noncommutative Lp-Spaces*,
Lecture Notes in Mathematics 2021, DOI 10.1007/978-3-642-20438-8_2,
© Springer-Verlag Berlin Heidelberg 2011

additional log-term is needed. By Theorem 7 we have that $S\Sigma D_{(1/\log n)}$ is (p,q)-maximizing whenever $q \geq p$. In this context a characterization of (p,q)-maximizing matrices in terms of p-summing and p-factorable operators (Theorems 3 and 4) in combination with Grothendieck's fundamental theorem of the metric theory of tensor products leads to a powerful link between the theory of general orthogonal series and its related L_p-theory (Theorem 5).

Let us once again mention that this first section was very much inspired by Bennett's seminal papers [2] and [3]. Finally, note that some of our proofs at a first glance may look cumbersome (see e.g. Lemma 2), but we hope to convince the reader that our special point of view later, in the noncommutative part of these notes, will be very helpful.

2.1.1 Summation of Scalar Series

For a scalar matrix $S = (s_{jk})_{j,k \in \mathbb{N}_0}$ with positive entries we call a scalar- or Banach space-valued sequence (x_k) S-summable whenever the sequence

$$\left(\sum_{k=0}^{\infty} s_{jk} \sum_{\ell=0}^{k} x_\ell \right)_j \tag{2.3}$$

of linear means of the partial sums of $\sum_k x_k$ (is defined and) converges. The matrix S is said to be a summation method or a summation process if for each convergent series $s = \sum_k x_k$ the sequence of linear means from (2.3) converges to s,

$$s = \lim_j \sum_{k=0}^{\infty} s_{jk} \sum_{\ell=0}^{k} x_\ell . \tag{2.4}$$

All results and examples we need on summation methods are contained in the monographs of Alexits [1] and Zygmund [98]. The following simple characterization of summation methods is due to Toeplitz [91].

Proposition 1. *Let $S = (s_{jk})$ be a scalar matrix with positive entries. Then S is a summation method if and only if*

(1) $\lim_j \sum_{k=0}^{\infty} s_{jk} = 1$
(2) $\lim_j s_{jk} = 0$ for all k

Moreover, for each Banach space X and each convergent series $s = \sum_k x_k$ in X we have (2.4), the limit taken in X.

Here we will only prove the fact that (1) and (2) are sufficient conditions for S to be a summation method, or more generally, that (1) and (2) imply (2.4) for every series $\sum_k x_k$ in a Banach space X (the necessity of (1) and (2) will not be needed in the following).

Proof. Take a series $s = \sum_{k=1}^{\infty} x_k$ in a Banach space X, and fix some $\varepsilon > 0$. Then there is k_0 such that we have $\left\| \sum_{\ell=0}^{k} x_\ell - s \right\| \leq \varepsilon$ for all $k \geq k_0$. Then for any j we have

$$\left\| \sum_{k=0}^{\infty} s_{jk} \sum_{\ell=0}^{k} x_\ell - s \right\| \leq \left\| \sum_{k=0}^{\infty} s_{jk} \left(\sum_{\ell=0}^{k} x_\ell - s \right) \right\| + \left\| s \sum_{k=0}^{\infty} s_{jk} - s \right\|$$

$$\leq \sum_{k=0}^{k_0} s_{jk} \left\| \sum_{\ell=0}^{k} x_\ell - s \right\| + \sum_{k=k_0+1}^{\infty} s_{jk} \left\| \sum_{\ell=0}^{k} x_\ell - s \right\| + \left\| s \sum_{k=0}^{\infty} s_{jk} - s \right\|$$

$$\leq \sum_{k=0}^{k_0} s_{jk} \left\| \sum_{\ell=0}^{k} x_\ell - s \right\| + \varepsilon \sum_{k=0}^{\infty} s_{jk} + \left\| s \sum_{k=0}^{\infty} s_{jk} - s \right\|,$$

and hence the conclusion follows from (1) and (2).   □

The following are our basic examples:

(1) The identity matrix $\mathrm{id} = (\delta_{jk})$ is trivially a summation method, and obviously (x_k) is summable if and only if it is id-summable.
(2) The matrix $C = (c_{jk})$ given by

$$c_{jk} := \begin{cases} \dfrac{1}{j+1} & k \leq j \\ 0 & k > j \end{cases}$$

is called Cesàro matrix, and for each series $\sum_k x_k$ (in a Banach space X)

$$\sum_{k=0}^{\infty} c_{jk} \sum_{\ell=0}^{k} x_\ell = \frac{1}{j+1} \sum_{k=0}^{j} \sum_{\ell=0}^{k} x_\ell$$

is its jth Cesàro mean. C-summable sequences are said to be Cesàro summable.
(3) For $r \in \mathbb{R}$ define $A_0^r = 1$, and for $n \in \mathbb{N}$

$$A_n^r := \binom{n+r}{n} = \frac{(r+1)\ldots(r+n)}{n!};$$

in particular, we have $A_n^1 = n+1$ and $A_n^0 = 1$. Then for $r > 0$ the matrix $C^r = (c_{jk}^r)$ defined by

$$c_{jk}^r := \begin{cases} \dfrac{A_{j-k}^{r-1}}{A_j^r} & k \leq j \\ 0 & k > j \end{cases}$$

is said to be the Cesàro matrix of order r. Obviously, we have that $C^1 = C$. All entries of C^r are positive, and on account of the well-known formula $\sum_{k=0}^{n} A_k^{r-1} = A_n^r$ and the fact that $A_n^r = O(n^r)$ (see also (2.44) and (2.48)) we have

$$\sum_{k=0}^{j} c_{jk}^r = 1 \quad \text{and} \quad c_{jk}^r \leq c\,\frac{(j-k)^{r-1}}{j^r}\,.$$

Hence, by the preceding proposition the matrices C^r form a scale of summation processes. Sequences which are C^r-summable are said to be Cesàro summable of order r.

(4) Let $(\lambda_k)_{k=0}^\infty$ be a strictly increasing sequence of positive scalars which converges to ∞, and such that $\lambda_0 = 0$. Then the so-called Riesz matrix R^λ defined by

$$r_{jk}^\lambda := \begin{cases} \dfrac{\lambda_{k+1} - \lambda_k}{\lambda_{j+1}} & k \leq j \\[2mm] 0 & k > j \end{cases}$$

forms a summation process; indeed

$$\sum_{k=0}^{j} \frac{\lambda_{k+1} - \lambda_k}{\lambda_{j+1}} = \frac{1}{\lambda_{j+1}}\,(\lambda_{j+1} - \lambda_0) = 1\,,$$

and

$$\lim_{j} \frac{\lambda_{k+1} - \lambda_k}{\lambda_{j+1}} = 0\,.$$

We call R^λ-summable sequences Riesz summable. Note that for $\lambda_j = j$ we have $R^\lambda = C$. Moreover, it is not difficult to see that for $\lambda = (2^j)$ Riesz-summation means nothing else than ordinary summation.

(5) Take a positive sequence (ρ_j) which increases to 1. Then the matrix A^ρ given by

$$a_{jk}^\rho := \rho_j^k (1 - \rho_j)$$

obviously defines a summation process. These matrices are called Abel matrices. Recall that a sequence (x_k) is said to be Abel summable whenever the limit

$$\lim_{r \to 1} \sum_{k=0}^{\infty} x_k r^k$$

exists. For $0 < r < 1$ we have

$$\sum_{k=0}^{\infty} x_k r^k = \sum_{k=0}^{\infty} r^k (1 - r) \sum_{\ell=0}^{k} x_\ell$$

which justifies our name for A^ρ.

2.1.2   Maximal Inequalities in Banach Function Spaces

As usual $L_p(\mu)$, $1 \le p \le \infty$ denotes the Banach space of all (equivalence classes of) p-integrable functions over a (in general σ-finite and complete) measure space (Ω, Σ, μ) (with the usual modification for $p = \infty$). We write $\ell_p(\Omega)$ whenever Ω is a set with the discrete measure, and ℓ_p for $\Omega = \mathbb{N}_0$ and ℓ_p^n for $\Omega = \{0, \ldots, n\}$. The canonical basis vectors are then denoted by e_i, $i \in \Omega$. More generally, we will consider Banach function spaces (sometimes also called Köthe function spaces) $E = E(\mu)$, i.e. Banach lattices of (μ-almost everywhere equivalence classes of) scalar-valued μ-locally integrable functions on Ω which satisfy the following two conditions:

- If $|x| \le |y|$ with $x \in L_0(\mu)$ and $y \in E(\mu)$, then $x \in E(\mu)$ and $\|x\| \le \|y\|$.
- For every $A \in \Sigma$ of finite measure the characteristic function χ_A belongs to $E(\mu)$.

Examples are L_p-, Orlicz, Lorentz, and Marcinkiewicz spaces.

Recall that a vector-valued function $f : \Omega \to X$, where X now is some Banach space, is μ-measurable whenever it is an almost everywhere limit of a sequence of vector-valued step functions. Then

$$E(X) = E(\mu, X)$$

consists of all (μ-equivalence classes of) μ-measurable functions $f : \Omega \to X$ such that $\|f\|_X \in E(\mu)$, a vector space which together with the norm

$$\|f\|_{E(\mu, X)} = \big\| \|f(\cdot)\|_X \big\|_{E(\mu)}$$

forms a Banach space. For $E(\mu) = L_p(\mu)$ this construction leads to the space $L_p(X) = L_p(\mu, X)$ of Bochner integrable functions; as usual $\ell_p(X)$ and $\ell_p^n(X)$ stand for the corresponding spaces of sequences in X.

We now invent two new spaces of families of integrable functions which will give a very comfortable setting to work with the maximal inequalities we are interested in. Let I be a partially ordered and countable index set, $E = E(\mu)$ a Banach function space, and X a Banach space. Then

$$E(X)[\ell_\infty] = E(\mu, X)[\ell_\infty(I)]$$

denotes the space of all families $(f_i)_{i \in I}$ in $E(\mu, X)$ having a maximal function which again belongs to $E(\mu)$,

$$\sup_{i \in I} \|f_i(\cdot)\|_X \in E(\mu).$$

Together with the norm

$$\|(f_i)\|_{E(X)[\ell_\infty]} := \Big\| \sup_{i \in I} \|f_i(\cdot)\|_X \Big\|_{E(\mu)}$$

$E(\mu,X)[\ell_\infty(I)]$ forms a Banach space. The following simple characterization will be extremely useful.

Lemma 1. *Let* $(f_i)_{i\in I}$ *be a family in* $E(\mu,X)$. *Then* $(f_i)_{i\in I}$ *belongs to* $E(\mu,X)[\ell_\infty(I)]$ *if and only if there is a bounded family* $(z_i)_{i\in I}$ *of functions in* $L_\infty(\mu,X)$ *and a scalar-valued function* $f \in E(\mu)$ *such that*

$$f_i = z_i f \text{ for all } i$$

(the pair $((z_i),f)$ *is then said to be a factorization of* (f_i)*). In this case, we have*

$$\|(f_i)\|_{E(X)[\ell_\infty]} = \inf \sup_{i\in I} \|z_i\|_\infty \|f\|_{E(\mu)},$$

the infimum taken over all possible factorizations.

For the sake of completeness we include the trivial

Proof. Let $(f_i) \in E(\mu,X)[\ell_\infty(I)]$. Put $f := \sup_i \|f_i(\cdot)\|_X \in E(\mu)$ and define $z_i(w) := f_i(w)/f(w)$ whenever $f(w) \neq 0$, and $z_i(w) := 0$ whenever $f(w) = 0$. Obviously, $f_i = z_i f$ and $\sup_i \|z_i\|_\infty \leq 1$, hence $\|f\|_{E(\mu)} \sup_i \|z_i\| \leq \|(f_i)\|_{E(X)[\ell_\infty]}$. Conversely, we have

$$\sup_i \|f_i(\cdot)\|_X \leq \sup_i \|z_i\|_\infty \|f(\cdot)\|_X \in E(\mu),$$

and hence

$$\|(f_i)\|_{E(X)[\ell_\infty]} \leq \sup_i \|z_i\|_\infty \|f\|_{E(\mu)},$$

which completes the argument.   □

We will also need the closed subspace

$$E(\mu,X)[c_0(I)] \subset E(\mu,X)[\ell_\infty(I)],$$

all families $(f_i) \in E(\mu,X)[\ell_\infty(I)]$ for which there is a factorization $f_i = z_i f$ with $\lim_i \|z_i\|_{L_\infty(X)} = 0$ and $f \in E(\mu)$; this notation seems now natural since we as usual denote the Banach space of all scalar zero sequences $(x_i)_{i\in I}$ by $c_0(I)$, and $c_0 = c_0(\mathbb{N}_0)$. The following lemma is a simple tool linking the maximal inequalities we are interested in with almost everywhere convergence.

Lemma 2. *Each family* $(f_i) \in E(\mu,X)[c_0(I)]$ *converges to 0* μ*-almost everywhere.*

Again we give the obvious

Proof. Let $f_i = z_i f$ be a factorization of (f_i) with $\lim_i \|z_i\|_{L_\infty(X)} = 0$ and $f \in E(\mu)$, and let (ε_i) be a zero sequence of positive scalars. Clearly, for each i there is a μ-null set N_i such that $\|z_i(\cdot)\|_X \leq \|z_i\|_{L_\infty(X)} + \varepsilon_i$ on the complement of N_i. Take an element w in the complement of the set $N := [|f| = \infty] \cup (\cup_i N_i)$. Then for $\varepsilon > 0$ there is i_0 such that $\|z_i\|_{L_\infty(X)} + \varepsilon_i \leq \frac{\varepsilon}{|f(w)|}$ for each $i \geq i_0$, and hence $|f_i(w)| = \|z_i(w)\|_X |f(w)| \leq (\|z_i\|_{L_\infty(X)} + \varepsilon_i)|f(w)| \leq \varepsilon$.   □

2.1.3 (p,q)-*Maximizing Matrices*

Recall that a sequence (x_k) in a Banach space X is said to be unconditionally summable (or equvialently, the series $\sum_k x_k$ is unconditionally convergent) whenever every rearrangement $\sum_k x_{\pi(k)}$ of the series converges. It is well-known that the vector space $\ell_1^{\mathrm{unc}}(X)$ of all unconditionally convergent series in X together with the norm

$$w_1((x_k)) := \sup_{\|\alpha\|_\infty \leq 1} \left\| \sum_{k=0}^\infty \alpha_k x_k \right\| < \infty.$$

forms a Banach space. More generally, for $1 \leq p \leq \infty$ a sequence (x_k) in a Banach space X is said to be weakly p-summable if for every $\alpha \in \ell_{p'}$ the series $\sum_k \alpha_k x_k$ converges in X, and by a closed graph argument it is equivalent to say that

$$w_p((x_k)) = w_p((x_k), X) := \sup_{\|\alpha\|_{p'} \leq 1} \left\| \sum_{k=0}^\infty \alpha_k x_k \right\| < \infty.$$

The name is justified by the fact that (x_k) is weakly p-summable if and only if $(x'(x_k)) \in \ell_p$ for each $x' \in X'$, and in this case we have

$$w_p((x_k)) = \sup_{\|x'\| \leq 1} \left(\sum_k |x'(x_k)|^p \right)^{\frac{1}{p}} < \infty.$$

The vector space of all weakly p-summable sequences in X together with the norm w_p forms the Banach space $\ell_p^w(X)$ (after the usual modification the case $p = \infty$ gives all bounded sequences). A sequence (x_k) is weakly summable (= weakly 1-summable) whenever the series $\sum_k x_k$ is unconditionally convergent, and the converse of this implication characterizes Banach spaces X which do not contain a copy of c_0. This is e.g. true for the spaces $L_p(\mu)$, $1 \leq p < \infty$.

The following definition is crucial – let $A = (a_{jk})_{j,k \in \mathbb{N}_0}$ be an infinite matrix which satisfies that $\|A\|_\infty := \sup_{jk} |a_{jk}| < \infty$, or equivalently, A defines a bounded and linear operator from ℓ_1 into ℓ_∞ with norm $\|A\|_\infty$.

Definition 1. We say that A is (p,q)-maximizing, $1 \leq p < \infty$ and $1 \leq q \leq \infty$, whenever for each measure space (Ω, μ), each weakly q'-summable sequence (x_k) in $L_p(\mu)$ and each $\alpha \in \ell_q$ we have that

$$\sup_j \left| \sum_{k=0}^\infty a_{jk} \alpha_k x_k \right| \in L_p(\mu),$$

or in other terms

$$\left(\sum_{k=0}^\infty a_{jk} \alpha_k x_k \right)_{j \in \mathbb{N}_0} \in L_p(\mu)[\ell_\infty].$$

Note that here all series $\sum_{k=0}^{\infty} a_{jk}\alpha_k x_k$ converge in $L_p(\mu)$. Clearly, by a closed graph argument a matrix A is (p,q)-maximizing if and only if the following maximal inequality holds: For all sequences (x_k) and (α_k) as above

$$\left\| \sup_j \left| \sum_{k=0}^{\infty} a_{jk}\alpha_k x_k \right| \right\|_p \le C\|\alpha\|_q w_{q'}((x_k));$$

here $C \ge 0$ is a constant which depends on A, p, q only, and the best of these constants is denoted by

$$\mathrm{m}_{p,q}(A) := \inf C.$$

Our main examples of maximizing matrices are generated by classical summation processes, and will be given in Sect. 2.2. Most of them are of the form

$$A = S\Sigma D_{1/\omega}, \quad a_{jk} := \frac{1}{\omega_k} \sum_{\ell=k}^{\infty} s_{j\ell}, \tag{2.5}$$

where S is a summation process as defined in Sect. 2.1.1, Σ is the so-called sum matrix defined by

$$\sigma_{jk} := \begin{cases} 1 & k \le j \\ 0 & k > j, \end{cases}$$

and $D_{1/\omega}$ a diagonal matrix with respect to a Weyl sequence ω for S (see again (2.2)). Since each such S can be viewed as an operator on ℓ_∞ (see Proposition 1,(1)), matrices of the form $S\Sigma D_{1/\omega}$ define operators from ℓ_1 into ℓ_∞.

Note that by definition such a matrix $A = S\Sigma D_{1/\omega}$ is (p,q)-maximizing whenever for each measure space (Ω, μ), each weakly q'-summable sequence (x_k) in $L_p(\mu)$ and each $\alpha \in \ell_q$ we have that

$$\sup_j \left| \sum_{k=0}^{\infty} s_{jk} \sum_{\ell=0}^{k} \frac{\alpha_\ell}{\omega_\ell} x_\ell \right| \in L_p(\mu), \tag{2.6}$$

or in other terms

$$\left(\sum_{k=0}^{\infty} s_{jk} \sum_{\ell=0}^{k} \frac{\alpha_\ell}{\omega_\ell} x_\ell \right)_j \in L_p(\mu)[\ell_\infty].$$

Let us once again repeat that by an obvious closed graph argument $A = S\Sigma D_{1/\omega}$ is (p,q)-maximizing if and only if for all sequences (x_k) and (α_k) as in (2.6) we have

$$\left\| \sup_j \left| \sum_{k=0}^{\infty} s_{jk} \sum_{\ell=0}^{k} \frac{\alpha_\ell}{\omega_\ell} x_\ell \right| \right\|_p \le C\|\alpha\|_q w_{q'}((x_k)),$$

$C \ge 0$ a constant which depends on A, p, q only.

It is not difficult to check (see also Sect. 2.2.6,(6)) that for the transposed A^t of an infinite matrix A the duality relation

$$m_{p,q}(A) = m_{q',p'}(A^t) \tag{2.7}$$

holds, and that $m_{p,q}(A)$ is decreasing in p and increasing in q, i.e. for $p_2 \leq p_1$ and $q_1 \leq q_2$

$$m_{p_1,q_1}(A) \leq m_{p_2,q_2}(A) \leq m_{1,\infty}(A) \tag{2.8}$$

(this will also be obtained as a by-product from Theorem 3). Finally, we include a simple lemma which helps to localize some of our coming arguments.

Lemma 3. *Let A be an infinite matrix with $\|A\|_\infty < \infty$, $E(\mu,X)$ a vector-valued Banach function space, and $1 \leq p < \infty$, $1 \leq q \leq \infty$. Then the following are equivalent:*

(1) For each $\alpha \in \ell_q$ and each weakly q'-summable sequence (x_k) in $E(\mu,X)$ we have that

$$\sup_j \left\| \sum_{k=0}^\infty a_{jk}\alpha_k x_k(\cdot) \right\|_X \in E(\mu).$$

(2) There is a constant $C > 0$ such that for each choice of finitely many scalars $\alpha_0,\ldots,\alpha_n$ and functions $x_0 \ldots, x_n \in E(\mu,X)$ we have

$$\left\| \sup_j \left\| \sum_{k=0}^n a_{jk}\alpha_k x_k(\cdot) \right\|_X \right\|_E \leq C \|\alpha\|_q w_{q'}(x).$$

In particular, A is (p,q)-maximizing if and only if $\sup_n m_{p,q}(A_n) < \infty$ where A_n equals A for all entries a_{jk} with $1 \leq j,k \leq n$ and is zero elsewhere; in this case

$$m_{p,q}(A) = \sup_n m_{p,q}(A_n).$$

Proof. Clearly, if (1) holds, then by a closed graph argument (2) is satisfied. Conversely, assume that (2) holds. First we consider the case $q < \infty$. Fix a weakly q'-summable sequence (x_k) in $E(\mu,X)$. By assumption we have

$$\sup_n \left\| \Phi_n : \ell_q^n \longrightarrow E(\mu,X)[\ell_\infty] \right\| = D < \infty,$$

where $\Phi_n \alpha := \left(\sum_k a_{jk}\alpha_k x_k \right)_j$. Hence, by continuous extension we find an operator $\Phi : \ell_q \to E(\mu,X)[\ell_\infty]$ of norm $\leq D$ which on all ℓ_q^n's coincides with Φ_n. On the other hand, since (x_k) is weakly q'-summable, the operator

$$\Psi : \ell_q \longrightarrow \prod_{\mathbb{N}_0} E(\mu,X), \quad \Psi(\alpha) = \left(\sum_k a_{jk}\alpha_k x_k \right)_j$$

is defined and continuous. Clearly, we have $\Psi = \Phi$ which concludes the proof. If $q = \infty$, then for fixed $\alpha \in \ell_\infty$ there is $D > 0$ such that for all n we have

$$\left\| \Phi_n : (\ell_1^n)^w (E(\mu,X)) \longrightarrow E(\mu,X)[\ell_\infty] \right\| \leq D,$$

where now $\Phi_n((x_k)) := \left(\sum_k a_{jk} \alpha_k x_k \right)_j$ (here $(\ell_1^n)^w(E(\mu,X))$ of course stands for the Banach space of all sequences of length $n+1$ endowed with the weak ℓ_1-norm w_1). Since the union of all $(\ell_1^n)^w(E(\mu,X))$ is dense in the Banach space $\ell_1^w(E(\mu,X))$, all weakly summable sequences (x_k) in $E(\mu,X)$, we can argue similarly to the first case. Finally, note that the last equality in the statement of the lemma follows from this proof.   □

The definition of (p,q)-maximizing matrices appears here the first time. But as we have already mentioned several times this notion is implicitly contained in Bennett's fundamental work on (p,q)-Schur multipliers from [3]; this will be outlined more carefully in Sect. 2.2.6.

2.1.4   Maximizing Matrices and Orthonormal Series

In this section we state our main technical tool to derive examples of (p,q)-maximizing matrices from classical coefficient tests on almost everywhere summation of orthonormal series and their related maximal inequalities (see (1.6) and (1.9)). This bridge is mainly based on dilation, a technique concentrated in the following lemma. Obviously, every orthonormal system in $L_2(\mu)$ is weakly 2-summable, but conversely each weakly 2-summable sequence is the "restriction" of an orthonormal system living on a larger measure space.

The following result due to Nagy is known under the name dilation lemma; for a proof see e.g. [94, Sect. III.H.19.]. It seems that in the context of almost everywhere convergence of orthogonal series this device was first used in Orno's paper [68].

Lemma 4. *Let (x_k) be a weakly 2-summable sequence in some $L_2(\Omega, \mu)$ with weakly 2-summable norm $w_2(x_k) \leq 1$. Then there is some measure space (Ω', μ') and an orthonormal system (y_k) in $L_2(\mu \oplus \mu')$ ($\mu \oplus \mu'$ the disjoint sum of both measures) such that each function x_k is the restriction of y_k.*

The following characterization of $(2,2)$-maximizing matrices in terms of orthonormal series is an easy consequence of this lemma.

Theorem 1. *Let $A = (a_{jk})$ be an infinite matrix such that $\|A\|_\infty < \infty$. Then A is $(2,2)$-maximizing if and only if for each $\alpha \in \ell_2$, for each measure μ and each orthonormal system (x_k) in $L_2(\mu)$*

$$\sup_j \left| \sum_k a_{jk} \alpha_k x_k \right| \in L_2(\mu). \tag{2.9}$$

Moreover, in this case $m_{2,2}(A)$ *equals the best constant C such that for each orthonormal series* $\sum_k \alpha_k x_k$ *in an arbitrary* $L_2(\mu)$

$$\left\| \sup_j \left| \sum_k a_{jk}\alpha_k x_k \right| \right\|_2 \leq C\|\alpha\|_2. \tag{2.10}$$

Proof. Clearly, if A is $(2,2)$-maximizing, then (2.9) holds and the infimum over all $C > 0$ as is (2.10) is $\leq m_{2,2}(A)$. Conversely, take $\alpha \in \ell_2$ and a weakly 2-summable sequence (y_k) in $L_2(\Omega, \mu)$; we assume without loss of generality that $w_2(y_k) \leq 1$. By the dilation lemma 4 there is some orthonormal system (x_k) in $L_2(\mu \oplus \mu')$ such that $x_k|\Omega = y_k$ for all k (μ' some measure on some measure space Ω'). We know by assumption that

$$\left(\sum_k a_{jk}\alpha_k x_k \right)_j \in L_2(\mu \oplus \mu')[\ell_\infty].$$

Hence by Lemma 1 there is a bounded sequence (z_j) in $L_\infty(\mu \oplus \mu')$ and some $f \in L_2(\mu \oplus \mu')$ for which $\sum_k a_{jk}\alpha_k x_k = z_j f$ for all j. But then as desired

$$\sup_j \left| \sum_k a_{jk}\alpha_k y_k \right| = \sup_j |z_j|_\Omega f|_\Omega| \in L_2(\mu).$$

If moreover the constant C satisfies (2.10), then we have

$$\left\| \sup_j \left| \sum_k a_{jk}\alpha_k y_k \right| \right\|_2 \leq \left\| \sup_j \left| \sum_k a_{jk}\alpha_k x_k \right| \right\|_2 \leq C\|\alpha\|_2,$$

hence $m_{2,2}(A) \leq C$.   □

2.1.5   *Maximizing Matrices and Summation: The Case* $q < p$

Recall that Σ denotes the sum matrix defined by

$$\sigma_{jk} := \begin{cases} 1 & k \leq j \\ 0 & k > j. \end{cases}$$

The study of (p,q)-maximizing matrices of type $S\Sigma$, where S is a summation process, shows two very different cases – the case $q < p$ and the case $p \leq q$. The next theorem handles the first one, for the second see Theorem 7.

Theorem 2. *Let* $1 \leq q < p < \infty$, *and let S be a summation process. Then the matrix* $A = S\Sigma$ *given by*

$$a_{jk} = \sum_{\ell=k}^{\infty} s_{j\ell}$$

is (p,q)-*maximizing.*

   This theorem is due to Bennett [2, Theorem 3.3] (only formulated for the crucial case, the sum matrix itself) who points out that the technique used for the proof goes back to Erdös' article [15].

Lemma 5. *Let* $1 < q < \infty$, *and assume that* $c_0, \cdots, c_n$ *are scalars such that*

$$|c_0|^q + \ldots + |c_n|^q = s > 0.$$

Then there is an integer $0 \leq k \leq n$ *such that*

$$|c_0|^q + \ldots + |c_{k-1}|^q + |c_k'|^q \leq s/2$$

$$|c_k''|^q + |c_{k+1}|^q + \ldots + |c_n|^q \leq s/2,$$

where $c_k = c_k' + c_k''$ *and* $\max\{|c_k'| \, |c_k''|\} \leq |c_k|$.

Proof. We start with a trivial observation: Take scalars c, d', d'' where d', d'' are positive and such that $d' \leq |c| \leq d' + d''$. Then there is a decomposition $c = c' + c''$ such that $|c'| \leq d'$ and $|c''| \leq d''$; indeed, decompose first the positive number $|c|$, and then look at the polar decomposition of c. Take now k such that

$$|c_0|^q + \ldots + |c_{k-1}|^q \leq s/2 < |c_0|^q + \ldots + |c_k|^q,$$

and define

$$d_k' := \left(s/2 - |c_0|^q - \ldots - |c_{k-1}|^q\right)^{1/q}$$

$$d_k'' := \left(|c_0|^q + \ldots + |c_k|^q - s/2\right)^{1/q} = \left(s/2 - |c_{k+1}|^q - \ldots - |c_n|^q\right)^{1/q}.$$

Since $q > 1$ we deduce from the starting observation that there is a decomposition $c_k = c_k' + c_k''$ with $|c_k'| \leq d_k' \leq |c_k|$ and $|c_k''| \leq d_k'' \leq |c_k|$ which completes the proof. $\qquad\square$

   Now we proceed with the proof of Theorem 2.

Proof. Let us first reduce the case of a general S to the special case $S = \mathrm{id}$: since S defines a bounded operator on ℓ_∞, we have that

$$\sup_j \left| \sum_{k=0}^\infty s_{jk} \sum_{\ell=0}^k \alpha_\ell x_\ell \right| \leq \|S : \ell_\infty \to \ell_\infty\| \sup_k \left| \sum_{\ell=0}^k \alpha_\ell x_\ell \right|, \qquad (2.11)$$

hence we only show that the matrix Σ is (p, q)-maximizing. We may assume that $1 < q < p < \infty$. By Lemma 3 it suffices to prove that there is a constant $c(p, q) > 0$ such that for each n

$$m_{p,q}(\Sigma_n) \leq c(p, q).$$

Fix n, and take $x_0, \ldots, x_n$ in some $L_p(\mu)$ with $w_{q'}(x_k) = 1$ and scalars $\alpha_0, \ldots, \alpha_n$ with $\|\alpha\|_q = 1$. We show that

$$\int \sup_j \Big| \sum_{k=0}^{j} \alpha_k x_k \Big|^p d\mu \leq c(p,q) .$$

To do so use the preceding lemma to split the sum

$$\alpha_0 x_0(\omega) + \ldots + \alpha_n x_n(\omega)$$

into two consecutive blocks

$$B_1^{(1)} = \alpha_0 x_0(\omega) + \ldots + \alpha_{k'} x_{k'}(\omega)$$

$$B_2^{(1)} = \alpha_{k''} x_{k''}(\omega) + \ldots + \alpha_n x_n(\omega)$$

such that each of the q-sums of the coefficients of these blocks is dominated by $1/2$ (split $\|\alpha\|_q^q = 1$). Applying the lemma we split each of the blocks into two further blocks $B_1^{(2)}, B_2^{(2)}$ and $B_3^{(2)}, B_4^{(2)}$, respectively. Repeating this process v times gives a decomposition of the original sum into 2^v blocks $B_\lambda^{(v)}$, $1 \leq \lambda \leq 2^v$, each having coefficient q-sums dominated by 2^{-v}. By choosing v sufficiently large, we may ensure that

$$2^{-v-1} < \min\{|\alpha_k| \mid \alpha_k \neq 0\} ,$$

so that each block $B_\lambda^{(v)}$ contains at most two non-zero terms (indeed, otherwise $2 \cdot 2^{-v-1} < 2^{-v}$). We then have for each $1 \leq j \leq n$ and all ω that

$$\Big| \sum_{k=0}^{j} \alpha_k x_k(\omega) \Big| \leq \sum_{\mu=1}^{v} \max_{1 \leq \lambda \leq 2^\mu} |B_\lambda^{(\mu)}(\omega)| + \max_{0 \leq k \leq n} |\alpha_k x_k(\omega)| .$$

Hence, for each r (which will be specified later) we obtain from Hölder's inequality that

$$\Big| \sum_{k=0}^{j} \alpha_k x_k(\omega) \Big|$$

$$\leq \sum_{\mu=1}^{v} \Big(\sum_{\lambda=1}^{2^\mu} |B_\lambda^{(\mu)}(\omega)|^p \Big)^{1/p} + \Big(\sum_{k=0}^{n} |\alpha_k x_k(\omega)|^p \Big)^{1/p}$$

$$\leq \Big(\sum_{\mu=1}^{v} 2^{-r\mu p'} \Big)^{1/p'} \Big(\sum_{\mu=1}^{v} 2^{r\mu p} \sum_{\lambda=1}^{2^\mu} |B_\lambda^{(\mu)}(\omega)|^p \Big)^{1/p} + \Big(\sum_{k=0}^{n} |\alpha_k x_k(\omega)|^p \Big)^{1/p} ,$$

and with $d(p,r) = \left(\sum_{\mu=1}^{\infty} 2^{-r\mu p'} \right)^{1/p'}$ we conclude

$$\left\| \sup_j \left| \sum_{k=0}^{j} \alpha_k x_k \right| \right\|_p$$

$$\leq d(p,r) \left(\sum_{\mu=1}^{v} 2^{r\mu p} \sum_{\lambda=1}^{2^\mu} \|B_\lambda^{(\mu)}\|_p^p \right)^{1/p} + \left(\sum_{k=0}^{n} |\alpha_k|^p \right)^{1/p};$$

use the Minkowski inequality in $L_p(\mu)$, the obvious fact that for each choice of finitely many functions $y_k \in L_p(\mu)$

$$\left\| \left(\sum_k |y_k|^p \right)^{1/p} \right\|_p = \left(\sum_k \|y_k\|_p^p \right)^{1/p},$$

and finally that all $\|x_k\|_p \leq 1$. By assumption we have that for every choice of finitely many scalars $\beta_0, \cdots, \beta_n$

$$\left\| \sum_k \beta_k x_k \right\|_p \leq \|(\beta_k)\|_q,$$

and that $1 \leq q < p < \infty$, hence

$$\left\| \sup_j \left| \sum_{k=0}^{j} \alpha_k x_k(\omega) \right| \right\|_p \leq d(p,r) \left(\sum_{\mu=1}^{v} 2^{r\mu p} \sum_{\lambda=1}^{2^\mu} 2^{-\mu p/q} \right)^{1/p} + \left(\sum_{k=0}^{n} |\alpha_k|^q \right)^{1/q}$$

$$\leq d(p,r) \left(\sum_{\mu=1}^{v} 2^{r\mu p} \sum_{\lambda=1}^{2^\mu} 2^{-\mu p/q} \right)^{1/p} + 1$$

$$\leq d(p,r) \left(\sum_{\mu=1}^{\infty} 2^{(rp+1-p/q)\mu} \right)^{1/p} + 1.$$

Since this latter term converges for each $0 < r < 1/q - 1/p$, the proof completes.

$\square$

As already mentioned, the counterpart of this result for $q \geq p$ will be stated in Theorem 7.

2.1.6 Banach Operator Ideals: A Repetitorium

A considerably large part for our conceptional approach to almost everywhere summation theorems of unconditionally convergent series in L_p-spaces together with their maximal inequalities will be based on the theory of Banach operator ideals.

We give, without any proofs, a brief summary of the results needed – in particular, we recall some of the ingredients from the theory of p-summing and p-factorable operators. Notes, remarks, and references are given at the end of this section.

An operator ideal $\mathscr{A}$ is a subclass of the class of all (bounded and linear) operators $\mathscr{L}$ between Banach spaces such that for all Banach spaces X and Y its components

$$\mathscr{A}(X,Y) := \mathscr{L}(X,Y) \cap \mathscr{A}$$

satisfy the following two conditions: $\mathscr{A}(X,Y)$ is a linear subspace of $\mathscr{L}(X,Y)$ which contains all finite rank operators, and for each choice of appropriate operators $u, w \in \mathscr{L}$ and $v \in \mathscr{A}$ we have $wvu \in \mathscr{A}$ (the ideal property). A (quasi) Banach operator ideal $(\mathscr{A}, \alpha)$ is an operator ideal $\mathscr{A}$ together with a function $\alpha : \mathscr{A} \longrightarrow \mathbb{R}_{\geq 0}$ such that every component $(\mathscr{A}(X,Y), \alpha(\cdot))$ is a (quasi) Banach space, $\alpha(\mathrm{id}_\mathbb{K}) = 1$, and for each choice of appropriate operators w, v, u we have that

$$\alpha(wvu) \leq \|w\| \alpha(v) \|u\|.$$

If $(\mathscr{A}, \alpha)$ is a Banach operator ideal, then it can be easily shown that

$$\|u\| \leq \alpha(u) \quad \text{for all } u \in \mathscr{A},$$

and for all one dimensional operators $x' \otimes y$ with $x' \in X', y \in Y$

$$\alpha(x' \otimes y) = \|x'\| \|y\|.$$

We will only consider maximal Banach operator ideals $(\mathscr{A}, \alpha)$, i.e. ideals which in the following sense are determined by their components on finite dimensional Banach spaces: An operator $u : X \longrightarrow Y$ belongs to $\mathscr{A}$ if (and only if)

$$\sup_{M,N} \alpha(M \xrightarrow{I_M} X \xrightarrow{u} Y \xrightarrow{Q_N} Y/N) < \infty, \tag{2.12}$$

where the supremum is taken over all finite dimensional subspaces M of X, all finite codimensional subspaces N of X and I_M, Q_N denote the canonical mappings. The duality theory of operator ideals is ruled by the following two notions, the trace tr for finite rank operators and the so-called adjoint operator ideals $\mathscr{A}^*$. If $(\mathscr{A}, \alpha)$ is a Banach operator ideal, then its adjoint ideal $(\mathscr{A}^*, \alpha^*)$ is given by: $u \in \mathscr{A}^*(X,Y)$ if

$$\alpha^*(u) := \sup_{M,N} \sup_{\|v:Y/N \to M\| \leq 1} \mathrm{tr}(Q_M u I_M v) < \infty$$

(M and N as above); note that this ideal by definition is maximal. If $(\mathscr{A}, \alpha)$ and $(\mathscr{B}, \beta)$ are two quasi Banach operator ideals, then $\mathscr{A} \circ \mathscr{B}$ denotes the operator ideal of all compositions $u = vw$ with $v \in \mathscr{A}$ and $w \in \mathscr{B}$, together with the quasi norm $\alpha \circ \beta(u) := \inf \alpha(v)\beta(w)$. This gives a quasi Banach operator ideal $(\mathscr{A} \circ \mathscr{B}, \alpha \circ \beta)$, the product of $(\mathscr{A}, \alpha)$ and $(\mathscr{B}, \beta)$. Let us finally recall the meaning of a transposed ideal $(\mathscr{A}^{\text{dual}}, \alpha^{\text{dual}})$: It consists of all $u \in \mathscr{L}$ such that its transposed $u' \in \mathscr{A}$, and $\alpha^{\text{dual}}(u) := \alpha(u')$.

Now we collect some of the most prominent examples of Banach operator ideals. Clearly, all operators on Banach spaces together with the operator norm $\|\cdot\|$ form the largest Banach operator ideal, here denoted by $\mathscr{L}$. The Banach ideal of p-summing operators is one of the fundamental tools of these notes. An operator $u : X \longrightarrow Y$ is said to be p-summing, $1 \le p < \infty$, whenever there is a constant $c \ge 0$ such that for all weakly p-summable sequences (x_k) in X we have

$$\left(\sum_{k=1}^{\infty} \|u(x_k)\|^p \right)^{\frac{1}{p}} \le c \sup_{\|x'\| \le 1} \left(\sum_{k=1}^{\infty} |x'(x_k)|^p \right)^{\frac{1}{p}} = w_p((x_k)), \qquad (2.13)$$

and the best constant c is denoted by $\pi_p(u)$. It can be seen easily that the class Π_p of all such operators together with the norm π_p forms a maximal Banach operator ideal (Π_∞ by definition equals $\mathscr{L}$).

There is also a non-discrete variant of (2.13): *An operator $u : X \longrightarrow Y$ is p-summing if and only if there is a constant $c \ge 0$ such that for any function $v \in L_p(\mu, X)$ (the Bochner p-integrable functions with values in X) we have*

$$\int \|u(v(\omega))\|^p d\mu(\omega) \le c \sup_{\|x'\| \le 1} \left(\int |x'(v(\omega))|^p d\mu(\omega) \right)^{\frac{1}{p}}, \qquad (2.14)$$

and in this case again the best c equals $\pi_p(u)$.

The whole theory of p-summing operators is ruled by Pietsch's domination theorem: *Let X and Y be Banach spaces, and assume that X is a subspace of some $C(K)$, where K is a compact Hausdorff space. Then $u : X \longrightarrow Y$ is p-summing if and only if there is a constant $c \ge 0$ and a Borel probability measure μ on K such that for all $x \in X$*

$$\|u(x)\| \le c \left(\int_K |x(w)|^p d\mu(\omega) \right)^{\frac{1}{p}}, \qquad (2.15)$$

and in this case the infimum over all possible c is a minimum and equals $\pi_p(u)$.

This result has many equivalent formulations in terms of factorization – we will need the following particular case: *For every p-summing operator $u : c_0 \longrightarrow Y$ there is a factorization*

$$(2.16)$$

with a diagonal operator D_α and an operator v satisfying $\|\alpha\|_p \|v\| \le \pi_p(u)$.

Finally, we mention two basic examples which in view of the preceding two
results are prototypical:

(1) $\pi_p(j : L_\infty(\mu) \hookrightarrow L_p(\mu)) = \mu(\Omega)$, where (Ω, μ) denotes some measure space
    and j the canonical embedding.
(2) $\pi_p(D_\alpha : c_0 \longrightarrow \ell_p) = \|\alpha\|_p$, where D_α denotes the diagonal operator associ-
    ated to $\alpha \in \ell_p$ (here c_0 can be replaced by ℓ_∞).

Let us now describe the adjoint ideal Π_p^* of Π_p in the more general context of
factorable operators. For $1 \le p \le q \le \infty$ denote by $\Gamma_{p,q}$ the Banach operator ideal of
all operators $u : X \longrightarrow Y$ which have a factorization

$$
\begin{array}{ccccc}
X & \xrightarrow{\ u\ } & Y & \xhookrightarrow{\ \kappa_Y\ } & Y'' \\[2pt]
\Big\downarrow{\scriptstyle v} & & & \nearrow{\scriptstyle w} & \\[2pt]
L_q(\mu) & \xhookrightarrow{\ j\ } & L_p(\mu) & &
\end{array}
\tag{2.17}
$$

where μ is a probability measure and v, w are two operators (clearly, κ_Y and j denote
the canonical embeddings). The ideal $\Gamma_{p,q}$ of all so-called (p,q)-factorable operators
together with the norm $\gamma_{p,q}(u) := \inf \|w\| \, \|v\|$ forms a maximal Banach operator
ideal. For operators $u : X \longrightarrow Y$ between finite dimensional spaces X and Y it can
be easily proved that

$$\gamma_{p,q}(u) = \inf \|w\| \, \|D_\mu\| \, \|v\|, \tag{2.18}$$

where "the infimum is taken over all possible diagrams" of the form

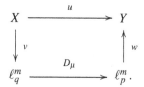

Define $\mathscr{I}_p := \Gamma_{p,\infty}$, the class of all p-integral operators, and $\Gamma_p := \Gamma_{p,p}$, the class of
all p-factorable operators; note that Γ_2 is the Banach operator ideal of all hilbertian
operators, all operators factorizing through a Hilbert space. Then (as a consequence
of Pietsch's domination theorem 2.15) *for operators u defined on $C(K)$-spaces or
with values in $C(K)$-spaces the p-integral and the p-summing norms coincide:*

$$\pi_p(u) = \iota_p(u). \tag{2.19}$$

Note that $(\mathscr{I}, \iota) := (\mathscr{I}_1, \iota_1)$ is the Banach operator ideal of all integral operators –
it is the smallest of all possible maximal Banach operator ideals, and moreover

it is the adjoint ideal of $(\mathscr{L}, \|\cdot\|)$. The following important trace formulas hold isometrically:

$$\mathscr{I}_p^* = \Pi_{p'}, \tag{2.20}$$

and more generally for $1 \le p \le q \le \infty$

$$\Gamma_{p,q}^* = \Pi_{q'}^{\text{dual}} \circ \Pi_{p'}. \tag{2.21}$$

As an easy consequence of the preceding equality the ideal of (p,q)-factorable operators can be rewritten as a sort of quotient of summing operators and integral operators — this "quotient formula" in the future will be absolutely crucial: *An operator $u : X \longrightarrow Y$ is (p,q)-factorable if and only if for each operator $v \in \Pi_q^{\text{dual}}(Z,X)$ the composition $uv \in \mathscr{I}_p(X,Y)$, and in this case*

$$\gamma_{p,q}(u) = \sup_{\pi_q(v') \le 1} \iota_p(uv). \tag{2.22}$$

Now we turn to tensor products – the theory of maximal Banach operator ideals and the theory of tensor products in Banach spaces are two in a sense equivalent languages. Recall that the projective norm $\|\cdot\|_\pi$ for an element z in the tensor product $X \otimes Y$ of two Banach spaces is given by

$$\|z\|_\pi = \inf \sum_k \|x_k\| \, \|y_k\|,$$

the infimum taken over all finite representation $z = \sum_k x_k \otimes y_k$. Dually, the injective norm $\|\cdot\|_\varepsilon$ for $z = \sum_k x_k \otimes y_k$ (a fixed finite representation) is defined by

$$\|z\|_\varepsilon = \sup_{\|x'\|_{X'}, \|y'\|_{Y'} \le 1} \Big| \sum_k x'(x_k) y'(y_k) \Big|.$$

We will need the simple fact: *For each integral operator $u \in \mathscr{L}(X,Y)$*

$$\iota(u) = \sup \| \text{id} \otimes u : Z \otimes_\varepsilon X \longrightarrow Z \otimes_\pi Y \|, \tag{2.23}$$

where the supremum is taken over all Banach spaces Z.

Let us finish with Grothendieck's *fundamental theorem of the metric theory of tensor products* (his théorème fondamental more or less in its original form) which is in a sense the hidden power in the background of most of the material following: *Every hilbertian operator $u : \ell_1 \longrightarrow \ell_\infty$ is integral, and*

$$\iota(u) = \pi_1(u) \le K_G \gamma_2(u), \tag{2.24}$$

where K_G is a universal constant (this best constant is usually called Grothendieck's constant).

An equivalent formulation of this highly non trivial fact is *Grothendieck's theorem* which states that *each operator $u : \ell_1 \to \ell_2$ is 1-summing, and $\pi_1(u) \leq K_G\|u\|$*. We will also need a weaker fact, the so called *little Grothendieck theorem: Every operator $u : \ell_1 \to \ell_2$ is 2-summing; in terms of tensor products this means that for each such u and each Hilbert space H we have*

$$\sup_n \|u \otimes \mathrm{id} : \ell_1^n \otimes_\varepsilon H \to \ell_2^n(H)\| \leq K_{LG}\|u\|, \tag{2.25}$$

and here (in contrast to Grothendieck's theorem) the precise constant $K_{LG} = 2/\sqrt{\pi}$ (the little Grothendieck constant) is known.

Notes and remarks: Most of the results presented in this section are standard, and can be found in the textbooks [6, 9, 76, 77], or [94]. The characterization of summing operators from (2.14) can be found in [94, Sect. III.F.33]. Pietsch's domination theorem (2.15) and factorization theorems like (2.16) are crucial, and contained in each of the above monographs. The trace duality theory of summing, integral and factorable operators is due to Kwapień, and at least for $p = q$ outlined in detail in the quoted textbooks; all needed properties of the ideal $\Gamma_{p,q}$ for $p \neq q$, in particular its relation with summing and integral norms like (2.19), (2.20), (2.21), and (2.22), are included in [6, Sects. 18, 25]. The estimate (2.24) is the main result in Grothendieck's famous "Résumé" [21] (the original source of all of this material), and together with (2.25) it forms one of the central topics in all monographs cited above.

2.1.7 Maximizing Matrices and Summation: The Case $q \geq p$

The following characterization of (p,q)-maximizing matrices links the classical theory of orthonormal series with modern operator theory in Banach spaces. Recall that by definition every (p,q)-maximizing matrix can be considered as an operator from ℓ_1 into ℓ_∞, and denote for $\alpha \in \ell_q$ by $D_\alpha : \ell_{q'} \to \ell_1$ the diagonal operator associated to α.

Theorem 3. *Let $1 \leq p < \infty$ and $1 \leq q \leq \infty$, and let A be an infinite matrix with $\|A\|_\infty < \infty$. Then the following are equivalent:*

(1) A is (p,q)-maximizing
(2) $\exists\, c \geq 0 \ \forall\, \alpha \in \ell_q : \pi_p(AD_\alpha) \leq c\,\|\alpha\|_q$
(3) $\exists\, c \geq 0 \ \forall n \ \forall\, u \in \mathscr{L}(\ell_{q'}^n, \ell_1) : \pi_p(Au) \leq c\,\pi_q(u')$
(4) $\exists\, c \geq 0 \ \forall$ Banach space $X \ \forall\, u \in \Pi_q^{dual}(X, \ell_1) : \pi_p(Au) \leq c\,\pi_q(u')$.

In this case, $m_{pq}(A) = \sup_{\|\alpha\|_q \leq 1} \pi_p(AD_\alpha) = \sup_{\pi_q(u') \leq 1} \pi_p(Au)$.

We try to make the proof a bit more transparent by proving a lemma first.

Lemma 6. *For every operator* $B : \ell_{q'} \longrightarrow \ell_\infty$ *the following are equivalent:*

(1) B is p-summing.

(2) $\exists\, c \geq 0 \,\forall\, x_0, \dots, x_m \in L_p(\mu) : \left\| \sup_j \left| \sum_{k=0}^m b_{jk} x_k \right| \right\|_p \leq c\, w_{q'}(x_k)$

In this case, $\pi_p(B) := \inf c$.

Proof. Let us first show that (1) implies (2). Take $x_0, \dots, x_m \in L_p(\mu)$. Then we obtain from (2.14) and the Bochner-integrable function

$$g := \sum_{k=0}^m x_k \otimes e_k \in L_p(\mu, \ell_{q'}^m)$$

that

$$\left\| \sup_j \left| \sum_{k=0}^m b_{jk} x_k \right| \right\|_p = \left(\int \|Bg\|_\infty^p d\mu \right)^{\frac{1}{p}}$$

$$\leq \pi_p(B) \sup_{\|x'\|_{\ell_q^m} \leq 1} \left(\int |x' \circ g|^p d\mu \right)^{\frac{1}{p}}$$

$$= \pi_p(B) \sup_{\|c\|_{\ell_q^m} \leq 1} \left\| \sum_{k=0}^m c_k x_k \right\|_p$$

$$= \pi_p(B)\, w_{q'}(x_k, L_p(\mu)).$$

Conversely, it suffices to show that for $x_0, \dots, x_m \in \ell_{q'}^M$

$$\left(\sum_{k=0}^m \|Bx_k\|_\infty \right)^{\frac{1}{p}} \leq c \sup_{\|x'\|_{\ell_q^M} \leq 1} \left(\sum_{k=0}^m |x'(x_k)|^p \right)^{\frac{1}{p}}.$$

Put $y_\ell := \sum_{n=0}^M x_n(\ell)e_n \in \ell_p^m$, $0 \leq \ell \leq M$. Then we have

$$\left\| \sup_j \left| \sum_{\ell=0}^M b_{j\ell} y_\ell \right| \right\|_{\ell_p^m} = \left(\sum_{k=0}^m \sup_j \left| \sum_{\ell=0}^M b_{j\ell} y_\ell(k) \right|^p \right)^{\frac{1}{p}}$$

$$= \left(\sum_{k=0}^m \sup_j \left| \sum_{\ell=0}^M b_{j\ell} \sum_{n=0}^m x_n(\ell)e_n(k) \right|^p \right)^{\frac{1}{p}}$$

$$= \left(\sum_{k=0}^m \|Bx_k\|_\infty^p \right)^{\frac{1}{p}}$$

and

$$\sup_{\|x'\|_{\ell_q^M}\leq 1} \left(\sum_{k=0}^{m}|x'(x_k)|^p\right)^{\frac{1}{p}} = \sup_{\|x'\|_{\ell_q^M}\leq 1}\ \sup_{\|d\|_{\ell_{p'}^m}\leq 1}\left|\sum_{k=0}^{m}d_k x'(x_k)\right|$$

$$= \sup_{\|d\|_{\ell_{p'}^m}\leq 1}\ \sup_{\|c\|_{\ell_q^M}\leq 1}\left|\sum_{\ell=0}^{M}c_\ell \sum_{k=0}^{m}d_k x_k(\ell)\right|$$

$$= \sup_{c}\sup_{d}\left|\sum_{k=0}^{m}d_k\sum_{\ell=0}^{M}c_\ell\sum_{n=0}^{m}x_n(\ell)e_n(k)\right|$$

$$= \sup_{c}\left\|\sum_{\ell=0}^{M}c_\ell\sum_{n=0}^{m}x_n(\ell)e_n\right\|_{\ell_p^m}$$

$$= \sup_{\|c\|_{\ell_q^M}\leq 1}\left\|\sum_{\ell=0}^{M}c_\ell y_\ell\right\|_{\ell_p^m} = w_{q'}\left(y_\ell,\ell_p^m\right).$$

Since we assume that (2) holds, these two equalities complete the proof.   □

Now we are prepared for the

Proof (of Theorem 3). First assume that A is (p,q)-maximizing, i.e. for every choice of a measure μ, a sequence $\alpha \in \ell_q$ and functions $x_0,\ldots,x_m \in L_p(\mu)$ we have

$$\left\|\sup_{j}\left|\sum_{k=0}^{j}a_{jk}\alpha_k x_k\right|\right\|_p \leq m_{p,q}(A)\|\alpha\|_q\, w_{q'}(x_k).$$

But then the preceding lemma implies that $AD_\alpha : \ell_{q'} \longrightarrow \ell_\infty$ is p-summing, and $\pi_p(AD_\alpha) \leq m_{p,q}(A)\|\alpha\|_q$. Conversely, assume that (2) holds. Then, again by the lemma,

$$\left\|\sup_{j}\left|\sum_{k=0}^{j}a_{jk}\alpha_k x_k\right|\right\|_p \leq c\|\alpha\|_q w_{q'}(x_k),$$

which yields (1). Next, we show that (2) implies (3). Take some $u \in \mathscr{L}(\ell_{q'}^n,\ell_1)$. Then by (2.16) there is a factorization with $\|D_\alpha\|\,\|R\| \leq \pi_q(u')$. But then (2) implies (3):

$$\pi_p(Au) = \pi_p(AD_\alpha R') \leq \pi_p(AD_\alpha)\|R\| \leq c\|\alpha\|_q\,\|R\| \leq c\,\pi_q(u')\,.$$

Now we prove the implication (3) $\Rightarrow$ (4): Recall that the Banach operator ideal (Π_p, π_p) is maximal (see (2.12)). Hence, we fix some operator $u : X \longrightarrow \ell_1$, and assume without loss of generality that X is finite dimensional. The aim is to show that

$$\pi_p(Au) \leq c\,\pi_q(u')\,.$$

It is well-known that there is some finite rank operator S on ℓ_1 such that $\|S\| \leq 1+\varepsilon$ and $S|_M = \mathrm{id}$ where $M := uX$ (ℓ_1 has the metric approximation property, see e.g. [6] or [53]). Put

$$v : X \longrightarrow M, \quad vx := Sux,$$

and let $I_M : M \hookrightarrow \ell_1$ be the canonical embedding. Without loss of generality there is a linear bijection $T : M \longrightarrow \ell_1^{\dim M}$ such that $\|T\|\,\|T^{-1}\| \leq 1+\varepsilon$ (ℓ_1 is a $\mathscr{L}_{1,\lambda}$-space, $\lambda > 1$; for this see again [6] or [53]). Again by (2.16) there is a factorization

$$\begin{array}{ccc} \ell_\infty^n & \xrightarrow{\ (Tu)'\ } & X' \\ & {\scriptstyle R}\searrow & \uparrow{\scriptstyle S} \\ & & \ell_q^N \end{array} \qquad\qquad \pi_q(R)\|S\| \leq \pi_q((Tu)')\,.$$

Hence, we conclude that

$$\begin{aligned} \pi_p(Au) &= \pi_p(AI_M u) \\ &= \pi_p(AI_M T^{-1} Tu) \\ &\leq \pi_p(AI_M T^{-1} R')\|S'\| \\ &\overset{(3)}{\leq} c\,\pi_q((I_M T^{-1} R')')\|S\| \\ &\leq c\,\pi_q(R)\|T^{-1}\|\,\|S\| \\ &\leq c\,\pi_q((Tu)')\|T^{-1}\| \leq c\,\pi_q(u')(1+\varepsilon)\,, \end{aligned}$$

the conclusion. This completes the whole proof since (4) trivially implies (2). $\qquad\square$

The preceding characterization has some deep consequences.

Theorem 4. *Let A be an infinite matrix such that $\|A\|_\infty < \infty$, and assume that $1 \leq p < \infty$, $1 \leq q \leq \infty$ with $p \leq q$.*

(1) A is (2,2)-maximizing if and only if $A : \ell_1 \longrightarrow \ell_\infty$ is hilbertian, and in this case $m_{2,2}(A) = \gamma_2(A)$.

(2) More generally, A is (p,q)-maximizing if and only if $A : \ell_1 \longrightarrow \ell_\infty$ is (p,q)-factorable,

$$
\begin{array}{ccc}
\ell_1 & \xrightarrow{\ A\ } & \ell_\infty \\
\downarrow{\scriptstyle v} & & \uparrow{\scriptstyle w} \\
L_q(\mu) & \overset{j}{\lhook\joinrel\longrightarrow} & L_p(\mu),
\end{array}
$$

and in this case $m_{p,q}(A) = \gamma_{p,q}(A)$.

(3) In particular, A is (p,∞)-maximizing if and only if $A : \ell_1 \longrightarrow \ell_\infty$ is p-summing (= p-integral by (2.19)), and in this case $m_{p,\infty}(A) = \pi_p(A)$.

Proof. It suffices to check (2) since (1) is an immediate consequence of (2), and (3) follows from (2) and (2.19). But (2) obviously is a consequence of the characterization of maximizing operators given in Theorem 3, (1) $\Leftrightarrow$ (4) combined with the quotient formula from (2.22) and the equality from (2.19). □

Note that (1) and (3) in combination with Grothendieck's théorème fondamental from (2.24) show that a matrix A is (2,2)-maximizing ($A : \ell_1 \to \ell_\infty$ is hilbertian) if and only A is $(1,\infty)$-maximizing ($A : \ell_1 \to \ell_\infty$ is integral). This is part of the following theorem which together with Theorem 1 is our second crucial tool later used to deduce a commutative and noncommutative L_p-theory of classical coefficient tests.

Theorem 5. *Let A be an infinite matrix such that $\|A\|_\infty < \infty$. The following are equivalent:*

(1) A is (2,2)-maximizing.
(2) A is $(1,\infty)$-maximizing.
(3) A is (p,q)-maximizing for some $1 \le p \le 2 \le q \le \infty$.
(4) A is (p,q)-maximizing for all $1 \le p < \infty$, $1 \le q \le \infty$.

In this case, $K_G^{-1} m_{1,\infty}(A) \le m_{2,2}(A) \le m_{1,\infty}(A)$.

Proof. We have already explained that the first two statements are equivalent. All other implications are then either trivial or follow by monotonicity. □

2.1.8   Almost Everywhere Summation

As anounced earlier one aim of this second chapter is to develop an L_p-theory for classical coefficient tests for almost sure summation of orthonormal series. The following theorem links the type of maximal inequalities in L_p-spaces we are interested in (i.e. inequalities induced by maximizing matrices) with almost everywhere convergence.

Proposition 2. *Let $A = (a_{jk})$ be a (p,q)-maximizing matrix which converges in each column, and $E(\mu,X)$ a vector-valued Banach function space. Then for every $\alpha \in \ell_q$ and every weakly q'-summable sequence (x_k) in $E(\mu,X)$ (in the case $q = \infty$ we only consider unconditionally summable sequences) the sequence*

$$\left(\sum_{k=0}^{\infty} a_{jk}\alpha_k x_k \right)_j$$

converges μ-almost everywhere.

Our proof will turn out to be a sort of model for the noncommutative case in Chap. 3; see Lemmas 22 and 27. That is the reason why we isolate the following lemma which here appears to be a bit too "heavy" – but obviously it allows to deduce the preceding proposition as an immediate consequence.

Lemma 7. *Let $A = (a_{jk})$ be a matrix with $\|A\|_{\infty} < \infty$ and such that each column forms a convergent sequence, $E(\mu,X)$ a vector-valued Banach function space, and $1 \leq q \leq \infty$. Assume that*

$$\left(\sum_{k=0}^{\infty} a_{jk}\alpha_k x_k \right)_j \in E(\mu,X)[\ell_{\infty}]$$

for every sequence $\alpha \in \ell_q$ and every weakly q'-summable sequence (x_k) in $E(\mu,X)$ (in the case $q = \infty$ we only consider unconditionally summable sequences). Then for every such α and (x_k) the sequence

$$\left(\sum_{k=0}^{\infty} a_{jk}\alpha_k x_k \right)_j$$

converges μ-almost everywhere.

Proof. We show that for every α and x as in the statement we have

$$\left(\sum_{k=0}^{\infty} a_{ik}\alpha_k x_k - \sum_{k=0}^{\infty} a_{jk}\alpha_k x_k \right)_{(i,j)} \in E(\mu,X)[c_0(\mathbb{N}_0^2)]; \qquad (2.26)$$

then we conclude from Lemma 2 that for each w in the complement of a zero set N

$$\lim_{(i,j) \to \infty} \left(\sum_k a_{ik}\alpha_k x_k(w) - \sum_k a_{jk}\alpha_k x_k(w) \right)_{(i,j)} = 0.$$

But this means that in $\complement N$ the sequence $\left(\sum_k a_{jk}\alpha_k x_k \right)_j$ is pointwise Cauchy, the conclusion.

In order to show (2.26) we first consider the case $1 \leq q < \infty$. Fix a weakly q'-summable sequence (x_k) in $E(\mu,X)$. Note first that for $(u_k) \in E(\mu,X)[\ell_\infty]$

$$(u_k - u_l)_{(k,l)} \in E(\mu,X)[\ell_\infty(\mathbb{N}_0^2)]$$

and

$$\left\| \sup_{k,l} \|u_k(\cdot) - u_l(\cdot)\|_X \right\|_{E(\mu)} \leq 2 \left\| \sup_k \|u_k(\cdot)\|_X \right\|_{E(\mu)};$$

this is obvious, but for later use in noncommutative settings let us also mention the following argument: if $u_k = z_k f$ is a factorization according to the definition of $E(\mu,X)[\ell_\infty]$, then

$$u_k - u_l = (z_k - z_l)f$$

defines a factorization for $(u_k - u_l)_{(k,l)}$. Hence by assumption the mapping

$$\Phi : \ell_q \longrightarrow E(\mu,X)[\ell_\infty(\mathbb{N}_0^2)]$$

$$\alpha \rightsquigarrow \left(\sum_k a_{ik}\alpha_k x_k - \sum_k a_{jk}\alpha_k x_k \right)_{(i,j)}$$

is defined, linear and (by a closed graph argument) bounded. Our aim is to show that Φ has values in the closed subspace $E(\mu,X)[c_0(\mathbb{N}_0^2)]$. By continuity it suffices to prove that, given a finite sequence $\alpha = (\alpha_0,\ldots,\alpha_{k_0},0,\ldots)$ of scalars, $\Phi\alpha \in E(\mu,X)[c_0(\mathbb{N}_0^2)]$. Clearly, $(\alpha_k x_k)_{0 \leq k \leq k_0} \in E(\mu,X)[\ell_\infty]$, and hence there is a factorization

$$\alpha_k x_k = z_k f, \quad 0 \leq k \leq k_0$$

with $\|z_k\|_{L_\infty(X)} \leq 1$ and $f \in E(\mu)$. But then for all i,j

$$\sum_{k=0}^{k_0} a_{ik}\alpha_k x_k - \sum_{k=0}^{k_0} a_{jk}\alpha_k x_k = \sum_{k=0}^{k_0} (a_{ik} - a_{jk})\alpha_k x_k = \left(\sum_{k=0}^{k_0} (a_{ik} - a_{jk})z_k \right) f.$$

This means that the right side of this equality defines a factorization of

$$\left(\sum_{k=0}^{k_0} a_{ik}\alpha_k x_k - \sum_{k=0}^{k_0} a_{jk}\alpha_k x_k \right)_{(i,j)}.$$

Since

$$\left\| \sum_{k=0}^{k_0} (a_{ik} - a_{jk})z_k \right\|_{L_\infty(X)} \leq \sum_{k=0}^{k_0} |a_{ik} - a_{jk}| \|z_k\|_{L_\infty(X)} \leq \sum_{k=0}^{k_0} |a_{ik} - a_{jk}|,$$

and A converges in each column, we even see that as desired

$$\left(\sum_k a_{ik}\alpha_k x_k - \sum_k a_{jk}\alpha_k x_k\right)_{(i,j)} \in E(\mu,X)[c_0(\mathbb{N}_0^2)].$$

For the remaining case $q = \infty$ fix $\alpha \in \ell_\infty$ and define

$$\Phi : \ell_1^{unc}(E(\mu)) \longrightarrow E[\ell_\infty(\mathbb{N}_0^2)]$$

$$(x_k) \rightsquigarrow \left(\sum_k a_{ik}\alpha_k x_k - \sum_k a_{jk}\alpha_k x_k\right)_{(i,j)}.$$

Like in the first case we see that Φ is well-defined and continuous. Since the finite sequences are dense in $\ell_1^{unc}(E(\mu,X))$, we can finish exactly as above.   □

2.2 Basic Examples of Maximizing Matrices

For some fundamental coefficient tests within the theory of pointwise summation of general orthogonal series with respect to classical summation methods, we isolate the maximal inequalities which come along with these results. In view of the results of the preceding section this leads to several interesting scales of (p,q)-maximizing matrices A – the main results are given in the Theorems 7 (ordinary summation), 8 (Riesz summation), 9 and 10 (Cesàro summation), 11 (Kronecker matrices), and 12 (Abel summation).

Let us once again repeat that most of our examples (but not all) have the form $A = (a_{jk})_{j,k\in\mathbb{N}_0} = S\Sigma D_{1/\omega}$, where S is some summation process (see (2.4)), $D_{1/\omega}$ some diagonal matrix with some Weyl sequence ω for S (see (2.2)), and Σ the sum matrix (see (2.1)):

$$a_{jk} := \frac{1}{\omega_k} \sum_{\ell=k}^{\infty} s_{j\ell}. \tag{2.27}$$

In the final section, we link our setting of maximizing matrices with Bennett's powerful theory of (p,q)-multipliers. We recall again that $\log x$ always means $\max\{1,\log x\}$.

2.2.1 The Sum Matrix

We already know from Theorem 2 that every matrix $S\Sigma$ is (p,q)-maximizing whenever $q < p$. The aim here is to prove the fundamental inequality of the theory of general orthonormal series – the famous Kantorovitch-Menchoff-Rademacher maximal inequality. This result will then show that every matrix of the form $S\Sigma D_{(1/\log k)}$ in fact is (p,q)-maximizing for arbitrary p,q.

Theorem 6. *Let* (x_k) *be an orthonormal system in* $L_2(\mu)$ *and* (α_k) *a scalar sequence satisfying* $\sum_{k=0}^{\infty} |\alpha_k \log k|^2 < \infty$. *Then the orthonormal series* $\sum_k \alpha_k x_k$ *converges almost everywhere, and its maximal function satisfies*

$$\left\| \sup_j \left| \sum_{k=0}^{j} \alpha_k x_k \right| \right\|_2 \leq C \| (\alpha_k \log k) \|_2, \tag{2.28}$$

where C is an absolute constant.

Improving many earlier results, the statement on almost everywhere convergence was independently discovered by Menchoff [60] and Rademacher [81], and today it is usually called Menchoff-Rademacher theorem (see e.g. [1, 47, 94]). Note that it is best possible in the following sense: Menchoff in [60] constructed an orthonormal system (x_k) such that for every increasing sequence (ω_k) in $\mathbb{R}_{\geq 1}$ with $\omega_k = o(\log k)$ there is an orthonormal series $\sum_k \alpha_k x_k$ which is divergent almost everywhere, but such that $\sum_{k=0}^{\infty} |\alpha_k \omega_k|^2 < \infty$. The maximal inequality (2.28) was isolated by Kantorovitch [46], and the result on almost everywhere convergence is clearly an easy consequence of it (see also Proposition 2). The optimality of the log-term in (2.28) can also be shown by use of the discrete Hilbert transform on ℓ_2 (see e.g. [2, 50, 59]).

The proof of the Kantorovitch-Menchoff-Rademacher maximal inequality (2.28) is done in two steps. First we show the following weaker estimate: Let $(\alpha_k)_{k=0}^{n}$ be scalars and $(x_k)_{k=0}^{n}$ an orthonormal system in $L_2(\mu)$. Then

$$\left\| \max_{0 \leq j \leq n} \left| \sum_{k=0}^{j} \alpha_k x_k \right| \right\|_2 \leq K \log n \| \alpha \|_2, \tag{2.29}$$

where $K > 0$ is an absolute constant.

Although the literature provides many elementary proofs of this inequality, we prefer to present a proof within our setting of maximizing matrices. In view of Theorem 1 the preceding estimate is equivalent to

$$m_{2,2}(\Sigma_n) \leq K \log n,$$

where Σ_n denotes the "finite" sum matrix

$$\sigma_{jk}^n := \begin{cases} 1 & k \leq j \leq n \\ 0 & j < k \leq n. \end{cases} \tag{2.30}$$

We show the apparently stronger (but by Theorem 5 equivalent) result

$$m_{1,\infty}(\Sigma_n) \leq K \log n, \tag{2.31}$$

which by Theorem 3, our general characterization of maximizing matrices through summing operators, is an immediate consequence of the following estimate.

Lemma 8. *There is a constant $K > 0$ such that for all n*

$$\pi_1(\Sigma_n : \ell_1^n \to \ell_\infty^n) \leq K \log n.$$

This lemma is well-known; see e.g. [2, 3] and [59]; the idea for the proof presented here is taken from[94, Sect. III.H.24]. For the estimate $\pi_1(\Sigma_n : \ell_1^n \to \ell_\infty^n) \leq \pi^{-1} \log n + O(1)$, where π^{-1} is optimal, see [3, Corollary 8.4].

Proof. Consider on the interval $[0, 2\pi]$ the matrix-valued function

$$A(\theta) := D(\theta)\left(e^{i(j-k)\theta}\right)_{jk},$$

where $D(\theta) = \sum_{j=0}^n e^{ij\theta}$ as usual denotes the Dirichlet kernel. Since we have that $A(\theta) = D(\theta) x \otimes y$ with $x = (e^{ij\theta})_j$ and $y = (e^{-ik\theta})_k$, the matrix $A(\theta)$ represents a one dimensional operator on $\mathbb{C}^n$. Hence

$$\pi_1(A(\theta) : \ell_1^n \longrightarrow \ell_\infty^n) = \|A(\theta) : \ell_1^n \longrightarrow \ell_\infty^n\| = |D(\theta)|,$$

and by the triangle inequality this implies that

$$\pi_1\left(\frac{1}{2\pi}\int_0^{2\pi} A(\theta)d\theta\right) \leq \frac{1}{2\pi}\int_0^{2\pi} |D(\theta)|d\theta \leq K \log n.$$

Since by coordinatewise integration we have

$$\Sigma_n = \frac{1}{2\pi}\int_0^{2\pi} A(\theta)d\theta,$$

the conclusion of the lemma follows.   □

Now we give the

Proof (of Theorem 6). It suffices to check the following two estimates:

$$\left\|\sup_n \left|\sum_{k=0}^{2^n} \alpha_k x_k\right|\right\|_2 \leq C_1 \|(\alpha_k \log k)\|_2, \tag{2.32}$$

$$\sum_n \left\|\max_{2^n < \ell \leq 2^{n+1}} \left|\sum_{k=0}^{\ell} \alpha_k x_k - \sum_{k=0}^{2^n} \alpha_k x_k\right|\right\|_2^2 \leq C_2 \|(\alpha_k \log k)\|_2^2; \tag{2.33}$$

indeed, for $2^m < j \leq 2^{m+1}$

$$\left| \sum_{k=0}^{j} \alpha_k x_k \right|^2 \le \left(\left| \sum_{k=0}^{2^m} \alpha_k x_k \right| + \left| \sum_{k=2^m+1}^{j} \alpha_k x_k \right| \right)^2$$

$$\le 2 \left(\left| \sum_{k=0}^{2^m} \alpha_k x_k \right|^2 + \left| \sum_{k=2^m+1}^{j} \alpha_k x_k \right|^2 \right)$$

$$\le 2 \left(\sup_n \left| \sum_{k=0}^{2^n} \alpha_k x_k \right|^2 + \sum_{n=0}^{\infty} \max_{2^n < \ell \le 2^{n+1}} \left| \sum_{k=2^n+1}^{\ell} \alpha_k x_k \right|^2 \right).$$

Hence we obtain by integration from (2.32) and (2.33) as desired

$$\left\| \sup_j \left| \sum_{k=0}^{j} \alpha_k x_k \right| \right\|_2 \le C \| (\alpha_k \log k) \|_2.$$

For the proof of (2.32) put $\varphi_0 := \sum_{k=0}^{2} \alpha_k x_k$ and $\varphi_v := \sum_{k=2^v+1}^{2^{v+1}} \alpha_k x_k$, $v \ge 1$. Since $v + 1 \le 2 \log(2^v)$, we have by orthogonality

$$\sum_{v=0}^{\infty} (v+1)^2 \| \varphi_v \|_2^2 = \| \varphi_0 \|_2^2 + \sum_{v=1}^{\infty} (v+1)^2 \sum_{k=2^v+1}^{2^{v+1}} |\alpha_k|^2$$

$$\le \| \varphi_0 \|_2^2 + 4 \sum_{v=1}^{\infty} \sum_{k=2^v+1}^{2^{v+1}} |\alpha_k \log k|^2 \le 4 \| (\alpha_k \log k) \|_2^2.$$

On the other hand $\sup_n \left| \sum_{k=0}^{2^n} \alpha_k x_k \right| \le \sum_{v=0}^{\infty} |\varphi_v|$ which now implies (2.32):

$$\left\| \sup_n \left| \sum_{k=0}^{2^n} \alpha_k x_k \right| \right\|_2 \le \sum_{v=0}^{\infty} \| \varphi_v \|_2 = \sum_{v=0}^{\infty} (v+1) \| \varphi_v \|_2 \frac{1}{v+1}$$

$$\le \left(\sum_{v=0}^{\infty} (v+1)^2 \| \varphi_v \|_2^2 \right)^{\frac{1}{2}} \left(\sum_{v=0}^{\infty} \frac{1}{(v+1)^2} \right)^{\frac{1}{2}}$$

$$\le C_1 \| (\alpha_k \log k) \|_2.$$

Finally, (2.33) is a consequence of (2.29): We have for all m

$$\left\| \max_{2^m < j \le 2^{m+1}} \left| \sum_{k=2^m+1}^{j} \alpha_k x_k \right| \right\|_2^2 \le C_2 (\log 2^m)^2 \sum_{k=2^m+1}^{2^{m+1}} |\alpha_k|^2,$$

and hence

$$\sum_{n=0}^{\infty} \left\| \max_{2^n < \ell \le 2^{n+1}} \left| \sum_{k=2^n+1}^{\ell} \alpha_k x_k \right| \right\|_2^2 \le C_2 \sum_{n=0}^{\infty} (\log 2^n)^2 \sum_{k=2^n+1}^{2^{n+1}} |\alpha_k|^2$$

$$\le C_2 \sum_{n=0}^{\infty} \sum_{k=2^n+1}^{2^{n+1}} |\alpha_k \log k|^2$$

$$\le C_2 \|(\alpha_k \log k)\|_2^2.$$

This completes the proof of Theorem 6.   □

Finally, we extend the preceding theorem within our setting of maximizing matrices. In combination with Theorem 1 we conclude from Theorem 6 that the matrix $A = \Sigma D_{(1/\log k)}$ given by

$$a_{jk} = \begin{cases} \dfrac{1}{\log k} & k \le j \\ 0 & k > j \end{cases} \tag{2.34}$$

is $(2,2)$-maximizing, hence Theorem 5 implies that this matrix is even (p,q)-maximizing for all p,q. The following formal extension of this statement complements Theorem 2.

Theorem 7. *Let S be a summation process, and $1 \le p < \infty$ and $1 \le q \le \infty$. Then the matrix $A = S\Sigma D_{(1/\log k)}$ given by*

$$a_{jk} = \frac{1}{\log k} \sum_{\ell=k}^{\infty} s_{j\ell}$$

is (p,q)-maximizing. Moreover, if $q < p$, then in the preceding statement no log-term is needed.

Proof. The matrix S defines a bounded operator on ℓ_∞. Hence we see from the argument already used in (2.11) and Theorem 6 that $S\Sigma D_{(1/\log k)}$ is $(2,2)$-maximizing, and therefore (p,q)-maximizing for all possible p,q by Theorem 5. The final statement is Theorem 2.   □

The following consequence of Theorem 4 is an interesting by-product on summing operators.

Corollary 1. *Let S be a summation process. Then the matrices $A = S\Sigma D_{(1/\log k)}$ from Theorem 7, if considered as operators from ℓ_1 into ℓ_∞, are 1-summing.*

We finish this section with a result on the "lacunarity" of the sum matrix Σ.

Corollary 2. *Take a strictly increasing unbounded sequence (λ_n) in $\mathbb{R}_{\ge 0}$ and let (ℓ_n) be its inverse sequence, i.e. if $\lambda : \mathbb{R}_{\ge 0} \to \mathbb{R}_{\ge 0}$ is linear in the interval $[n, n+1]$ and $\lambda(n) := \lambda_n$, then $\ell_n := \ell(n)$ with $\ell := \lambda^{-1}$. Let $\Sigma^o = (\sigma_{jk}^o)$ be an infinite matrix* *which equals the sum matrix Σ except that some columns are entirely zero. If for all n we have*

$$\text{card}\{k \,|\, \ell_n \leq k \leq \ell_{n+1}, \text{ the k-th column of } \Sigma^o \text{ is non-vanishing}\} \leq O(n),$$

then the matrix A defined by

$$a_{jk} := \begin{cases} \dfrac{\sigma^o_{jk}}{\log \lambda(k)} & k \leq j \\ 0 & k > j, \end{cases}$$

is (p,q)-maximizing for all p,q.

Proof. By Theorem 5 we only have to show that A is $(2,2)$-maximal, and hence by Theorem 1 we check that for a given orthonormal series $\sum_k \alpha_k x_k$ in $L_2(\mu)$

$$\left\| \sup_j \left| \sum_k \sigma^o_{jk} \alpha_k x_k \right| \right\|_2 \leq C \|(\alpha_k \log \lambda_k)\|_2.$$

The proof is based on the Kantorovitch-Menchoff-Rademacher inequality (2.28), but it also repeats part of its proof. As there, it suffices to check that the sequence of partial sums

$$s_j = \sum_{k=0}^j \sigma^o_{jk} \alpha_k x_k, \; j \in \mathbb{N}$$

satisfies the following two inequalities:

$$\sum_n \left\| \max_{\ell_n < \ell \leq \ell_{n+1}} |s_\ell - s_{\ell_n}| \right\|_2^2 \leq C \|(\alpha_k \log \lambda_k)\|_2^2 \tag{2.35}$$

$$\left\| \sup_n |s_{\ell_n}| \right\|_2 \leq C \|(\alpha_k \log \lambda_k)\|_2, \tag{2.36}$$

$C \geq 1$ some constant. We assume without loss of generality that all ℓ_n are natural numbers and all scalars $\beta_n := \left(\sum_{k=\ell_n+1}^{\ell_{n+1}} |\sigma^o_{\ell_n+1\,k} \alpha_k|^2 \right)^{1/2} \neq 0$. Define the orthonormal system

$$y_n := \frac{1}{\beta_n} \sum_{k=\ell_n+1}^{\ell_{n+1}} \sigma^o_{\ell_n+1\,k} \alpha_k x_k, \; n \in \mathbb{N}.$$

Then we obtain (2.36) from (2.28) (note that $\sigma^o_{\ell_n+1\,\ell} = \sigma^o_{\ell_{k+1}\,\ell}$ for $\ell \leq \ell_{k+1}$):

$$\left\| \sup_n \left| \sum_{k=0}^{\ell_{n+1}} \sigma^o_{\ell_n+1\,k} \alpha_k x_k \right| \right\|_2 = \left\| \sup_n \left| \sum_{k=0}^n \beta_k y_k \right| \right\|_2$$

$$\leq C \left(\sum_{k=0}^\infty \log^2 k \, \beta_k^2 \right)^{1/2}$$

$$\leq C \Big(\sum_{k=0}^{\infty} \log^2 k \sum_{\ell=\ell_k+1}^{\ell_{k+1}} |\alpha_\ell|^2 \Big)^{1/2}$$

$$\leq C \| (\alpha_k \log \lambda_k) \|_2 .$$

Moreover, from another application of (2.28) (more precisely, the weaker estimate from (2.29)) and the hypothesis on the number of nonzero columns of A we conclude

$$\Big\| \max_{\ell_n < \ell \leq \ell_{n+1}} \Big| \sum_{k=\ell_n+1}^{\ell} \sigma_{\ell k}^o \alpha_k x_k \Big\|_2^2 \leq C \log^2 n \sum_{\ell=\ell_n+1}^{\ell_{n+1}} |\alpha_\ell|^2$$

$$\leq C \sum_{k=\ell_n+1}^{\ell_{n+1}} |\alpha_k \log \lambda_k|^2 ,$$

which after summation over all n yields (2.35). $\square$

2.2.2 Riesz Matrices

For particular summation methods S the log-term in Theorem 7 can be improved. In the following section we handle Riesz matrices $S = R^\lambda$; recall their definition from Sect. 2.1.1.

Theorem 8. *Let $(\lambda_n)_n$ be a strictly increasing unbounded sequence of positive numbers with $\lambda_0 = 0$, and $1 \leq p < \infty, 1 \leq q \leq \infty$. Then the matrix $A = R^\lambda \Sigma D_{(1/\log\log\lambda_k)}$ given by*

$$a_{jk} = \begin{cases} (1 - \dfrac{\lambda_k}{\lambda_{j+1}}) \dfrac{1}{\log\log\lambda_k} & k \leq j \\ 0 & k > j \end{cases}$$

is (p,q)-maximizing. No log-term is needed whenever $q < p$.

Recall that we agreed to write $\log \lambda_k$ for $\max\{1, \log \lambda_k\}$. Clearly, the last statement on the log-term is a consequence of Theorem 2. Moreover, note that for the special case $\lambda_n = 2^n$ this result still contains the Kantorovitch-Mechoff-Rademacher inequality (2.28) together with its (p,q)-variants. From Theorem 4 we deduce the following immediate

Corollary 3. *All matrices $A = R^\lambda \Sigma D_{(1/\log\log\lambda_k)}$ from the preceding Theorem 8, if considered as operators from ℓ_1 into ℓ_∞, are 1-summing.*

Theorem 8 is due to Bennett [2, Theorem 6.5] who gives a direct, may be more elementary proof for it, and Corollary 3 was first stated in [3, Corollary 6.4]. Before we enter the proof of Theorem 8 let us recall what it means in terms of a maximal inequality: There is a constant $C > 0$ such that for each sequence (α_k) of scalars and for each sequence (x_k) in $L_p(\mu)$

$$\left\| \sup_j \left| \sum_{k=0}^{j} \frac{\lambda_{k+1} - \lambda_k}{\lambda_{j+1}} \sum_{\ell=0}^{k} \alpha_\ell x_\ell \right| \right\|_p \leq C \|(\alpha_k \log\log \lambda_k)\|_q \, w_{q'}(x_k) . \qquad (2.37)$$

In order to prove this inequality we have to check, by what was shown in Theorem 1 and Theorem 5, an a priori weaker estimate for orthonormal series. It suffices to prove that for each orthonormal series $\sum_k \alpha_k x_k$ in $L_2(\mu)$ we have

$$\left\| \sup_j \left| \sum_{k=0}^{j} \frac{\lambda_{k+1} - \lambda_k}{\lambda_{j+1}} \sum_{\ell=0}^{k} \alpha_\ell x_\ell \right| \right\|_2 \leq C \|(\alpha_k \log\log \lambda_k)\|_2 , \qquad (2.38)$$

$C > 0$ some universal constant. This maximal inequality for orthonormal series corresponds to a famous almost everywhere summation theorem due to Zygmund [97]; our proof follows from a careful analysis of the proof of Zygmund's result given in Alexits [1, p.141], and it is based on the Kantorovitch-Menchoff-Rademacher inequality (2.28).

Proof (of (2.38)). Define

$$s_j = \sum_{k=0}^{j} \alpha_k x_k \quad \text{and} \quad \sigma_j = \sum_{k=0}^{j} \left(1 - \frac{\lambda_k}{\lambda_{j+1}} \right) \alpha_k x_k .$$

By assumption there is a strictly increasing function $\lambda : \mathbb{R}_{\geq 0} \to \mathbb{R}_{\geq 0}$ being linear in each interval $[n, n+1]$ and satisfying $\lambda(n) = \lambda_n$ for all n. Put $v_n := l(2^n)$, where $l : \mathbb{R}_{\geq 0} \to \mathbb{R}_{\geq 0}$ is the inverse function of λ; we assume that all v_n's are integers (otherwise the proof needs some modifications). It suffices to check the following three estimates:

$$\sum_n \| s_{v_n} - \sigma_{v_n} \|_2^2 \leq C_1 \|\alpha\|_2^2 \qquad (2.39)$$

$$\sum_n \left\| \max_{v_n < \ell \leq v_{n+1}} |\sigma_\ell - \sigma_{v_n}| \right\|_2^2 \leq C_2 \|\alpha\|_2^2 \qquad (2.40)$$

$$\left\| \sup_n |s_{v_n}| \right\|_2 \leq C_3 \|(\alpha_k \log\log \lambda_k)_k\|_2 ; \qquad (2.41)$$

indeed, for $v_m < j \leq v_{m+1}$

$$|\sigma_j|^2 \leq (|\sigma_j - \sigma_{v_m}| + |\sigma_{v_m} - s_{v_m}| + |s_{v_m}|)^2$$

$$\leq 3(|\sigma_j - \sigma_{v_m}|^2 + |\sigma_{v_m} - s_{v_m}|^2 + |s_{v_m}|^2)$$

$$\leq 3\left(\sum_{n=0}^{\infty} \max_{v_n < \ell \leq v_{n+1}} |\sigma_\ell - \sigma_{v_n}|^2 + \sum_{n=0}^{\infty} |\sigma_{v_n} - s_{v_n}|^2 + \sup_n |s_{v_n}|^2 \right),$$

and since the right side is independent of j, we have

$$\sup_j |\sigma_j|^2 \le 3 \left(\sum_{n=0}^{\infty} \max_{v_n < \ell \le v_{n+1}} |\sigma_\ell - \sigma_{v_n}|^2 + \sum_{n=0}^{\infty} |\sigma_{v_n} - s_{v_n}|^2 + \sup_n |s_{v_n}|^2 \right).$$

Hence, we obtain as desired

$$\left\| \sup_j |\sigma_j| \right\|_2^2 \le 3 \left(\sum_{n=0}^{\infty} \left\| \max_{v_n < \ell \le v_{n+1}} |\sigma_\ell - \sigma_{v_n}| \right\|_2^2 + \sum_{n=0}^{\infty} \left\| \sigma_{v_n} - s_{v_n} \right\|_2^2 + \left\| \sup_n |s_{v_n}| \right\|_2^2 \right)$$

$$\le 3 \left(C_2 \|\alpha\|_2^2 + C_1 \|\alpha\|_2^2 + C_3 \|(\alpha_k \log\log \lambda_k)_k\|_2^2 \right) \le C \|(\alpha_k \log\log \lambda_k)\|_2^2.$$

For the proof of (2.39) note that $v_n = l(2^n) \ge k = \lambda^{-1}(\lambda_k)$ iff $n \ge \log \lambda_k$. Therefore, (2.39) is obtained by orthogonality as follows:

$$\sum_{n=0}^{\infty} \| s_{v_n} - \sigma_{v_n} \|_2^2 = \sum_{n=0}^{\infty} \left\| \sum_{k=0}^{v_n} \frac{\lambda_k}{\lambda_{v_n+1}} \alpha_k x_k \right\|_2^2$$

$$= \sum_{n=0}^{\infty} \sum_{k=0}^{v_n} \left(\frac{\lambda_k}{\lambda_{v_n+1}} \right)^2 |\alpha_k|^2$$

$$\le \sum_{k=0}^{\infty} |\alpha_k|^2 \sum_{n: v_n \ge k} \left(\frac{\lambda_k}{2^n} \right)^2$$

$$= \sum_{k=0}^{\infty} |\alpha_k|^2 \sum_{n \ge \log \lambda_k} \left(\frac{1}{2^{n - \log \lambda_k}} \right)^2 \le C_1 \|\alpha_k\|_2^2.$$

In order to show (2.40) choose for a fixed m some n such that $v_n < m \le v_{n+1}$. Then

$$|\sigma_m - \sigma_{v_n}| \le \sum_{j=v_n}^{m} |\sigma_{j+1} - \sigma_j| \le \sum_{j=v_n}^{v_{n+1}} |\sigma_{j+1} - \sigma_j|$$

$$= \sum_{j=v_n}^{v_{n+1}} \left(\frac{\lambda_{j+1}}{\lambda_{j+2} - \lambda_{j+1}} \right)^{\frac{1}{2}} |\sigma_{j+1} - \sigma_j| \left(\frac{\lambda_{j+2} - \lambda_{j+1}}{\lambda_{j+1}} \right)^{\frac{1}{2}}$$

$$\le \left(\sum_{j=v_n}^{v_{n+1}} \frac{\lambda_{j+1}}{\lambda_{j+2} - \lambda_{j+1}} |\sigma_{j+1} - \sigma_j|^2 \right)^{\frac{1}{2}} \left(\sum_{j=v_n}^{v_{n+1}} \frac{\lambda_{j+2} - \lambda_{j+1}}{\lambda_{j+1}} \right)^{\frac{1}{2}}.$$

But

$$\sum_{j=v_n}^{v_{n+1}} \frac{\lambda_{j+2} - \lambda_{j+1}}{\lambda_{j+1}} \le \frac{1}{\lambda_{v_n}} \sum_{j=v_n}^{v_{n+1}} \lambda_{j+2} - \lambda_{j+1} \le \frac{1}{2^n} (2^{n+3} - 2^n) = 7$$

and by orthogonality

$$\int |\sigma_{j+1} - \sigma_j|^2 d\mu = \int \left| \sum_{k=0}^{j+1} \left(1 - \frac{\lambda_k}{\lambda_{j+2}} \right) \alpha_k x_k - \sum_{k=0}^{j} \left(1 - \frac{\lambda_k}{\lambda_{j+1}} \right) \alpha_k x_k \right|^2 d\mu$$

$$= \left(\frac{\lambda_{j+2} - \lambda_{j+1}}{\lambda_{j+1}\lambda_{j+2}} \right)^2 \int \left| \sum_{k=0}^{j+1} \lambda_k \alpha_k x_k \right|^2 d\mu$$

$$= \left(\frac{\lambda_{j+2} - \lambda_{j+1}}{\lambda_{j+1}\lambda_{j+2}} \right)^2 \sum_{k=0}^{j+1} \lambda_k^2 |\alpha_k|^2 .$$

Hence

$$\sum_{n=0}^{\infty} \int \max_{v_n < \ell \le v_{n+1}} |\sigma_m - \sigma_{v_n}|^2 d\mu \le 7 \sum_{j=0}^{\infty} \frac{\lambda_{j+1}}{\lambda_{j+2} - \lambda_{j+1}} \int |\sigma_{j+1} - \sigma_j|^2 d\mu$$

$$= 7 \sum_{j=0}^{\infty} \frac{\lambda_{j+2} - \lambda_{j+1}}{\lambda_{j+1}\lambda_{j+2}^2} \sum_{k=0}^{j+1} \lambda_k^2 |\alpha_k|^2$$

$$= 7 \sum_{k=0}^{\infty} \lambda_k^2 |\alpha_k|^2 \sum_{j=k-1}^{\infty} \frac{\lambda_{j+2} - \lambda_{j+1}}{\lambda_{j+1}\lambda_{j+2}^2} .$$

But since

$$\frac{\lambda_{j+2} - \lambda_{j+1}}{\lambda_{j+1}\lambda_{j+2}} = \frac{\lambda_{j+2}^2 - \lambda_{j+1}^2}{\lambda_{j+1}\lambda_{j+2}^2(\lambda_{j+2} + \lambda_{j+1})} \le \frac{\lambda_{j+2}^2 - \lambda_{j+1}^2}{\lambda_{j+1}^2\lambda_{j+2}^2} = \frac{1}{\lambda_{j+1}^2} - \frac{1}{\lambda_{j+2}^2},$$

we now obtain (2.40):

$$\sum_{n=0}^{\infty} \left\| \max_{v_n < \ell \le v_{n+1}} |\sigma_\ell - \sigma_{v_n}| \right\|_2^2 \le 7 \sum_{k=0}^{\infty} \lambda_k^2 |\alpha_k|^2 \sum_{j=k-1}^{\infty} \frac{1}{\lambda_{j+1}^2} - \frac{1}{\lambda_{j+2}^2} \le C_2 \|\alpha\|_2^2 .$$

Finally, the proof of (2.41): We may assume without loss of generality that all

$$\beta_n := \begin{cases} \left(\sum_{k=0}^{v_1} |\alpha_k|^2 \right)^{\frac{1}{2}} & n = 0 \\ \left(\sum_{k=v_n+1}^{v_{n+1}} |\alpha_k|^2 \right)^{\frac{1}{2}} & n \ge 1 \end{cases}$$

are $\ne 0$. Then the functions

$$y_n := \begin{cases} \frac{1}{\beta_0} \sum_{k=0}^{v_1} \alpha_k x_k & n = 0 \\ \frac{1}{\beta_n} \sum_{k=v_n+1}^{v_{n+1}} \alpha_k x_k & n \ge 1 \end{cases}$$

define an orthonormal system in $L_2(\mu)$. From $\log n = \log \log 2^n = \log \log \lambda_{v_n}$ we derive

$$
\sum_{n=0}^{\infty} |\beta_n \log n|^2 = \sum_{k=0}^{v_1} |\alpha_k|^2 + \sum_{n=1}^{\infty} (\log n)^2 \sum_{k=v_n+1}^{v_{n+1}} |\alpha_k|^2
$$

$$
\leq \sum_{k=0}^{v_1} |\alpha_k \log \log \lambda_k|^2 + \sum_{n=1}^{\infty} \sum_{k=v_n+1}^{v_{n+1}} |\alpha_k \log \log \lambda_k|^2
$$

$$
= \sum_{k=0}^{\infty} |\alpha_k \log \log \lambda_k|^2 ,
$$

which in combination with Theorem 6 gives as desired

$$
\left\| \sup_n |s_{v_{n+1}}| \right\|_2 = \left\| \sup_n \left| \sum_{k=0}^{n} \beta_k y_k \right| \right\|_2
$$

$$
\leq C_3 \| (\beta_k \log k) \|_2^2 = C_3 \| (\alpha_k \log \log \lambda_k) \|_2^2 .
$$

This completes the proof.　　　　　　　　　　　　　　　　　　　　　　□

2.2.3　Cesàro Matrices

We deal with Cesàro matrices C^r defined in Sect. 2.1.1. Note first that for $\lambda_n = n$ Theorem 8 reads as follows.

Theorem 9. *The matrix* $A = C\Sigma D_{(1/\log\log k)}$ *given by*

$$
a_{jk} = \begin{cases} \left(1 - \dfrac{k}{j+1}\right) \dfrac{1}{\log \log k} & k \leq j \\ 0 & k > j \end{cases}
\tag{2.42}
$$

is (p,q)-*maximizing for* $1 \leq p < \infty$ *and* $1 \leq q \leq \infty$. *No* log-*term is needed whenever* $q < p$.

We will now extend this result for Cesàro matrices C^r of order $r > 0$. For all needed facts on Cesàro summation of order r we once more refer to the monographs [1] and [97]. For $r \in \mathbb{R}$ define $A_0^r = 1$, and for $n \in \mathbb{N}$

$$
A_n^r := \binom{n+r}{n} = \frac{(r+1)\ldots(r+n)}{n!} ;
$$

recall that these numbers are the coefficients of the binomial series

$$\sum_{n=0}^{\infty} A_n^r z^n = \frac{1}{(1-z)^{r+1}} \tag{2.43}$$

($z \in \mathbb{C}$ with $|z| < 1$). In particular, we see that the equality

$$\sum_{n=0}^{\infty} A_n^{r_1+r_2+1} z^n = (1-z)^{-r_1-1}(1-z)^{-r_2-1} = \sum_{n=0}^{\infty} \left(\sum_{k=0}^{n} A_k^{r_1} A_{n-k}^{r_2} \right) z^n$$

implies the formulas

$$A_n^{r_1+r_2+1} = \sum_{k=0}^{n} A_k^{r_1} A_{n-k}^{r_2} . \tag{2.44}$$

For a sequence (x_k) in a Banach space and $r \in \mathbb{R}$ define the Cesàro means

$$s_j^r = \sum_{k=0}^{j} A_{j-k}^{r-1} s_k \quad \text{and} \quad \sigma_j^r = \frac{1}{A_j^r} s_j^r ,$$

where s_k again is the kth partial sum of the series $\sum_k x_k$. Using that $\sum_{k=0}^{j} A_k^{r-1} = A_j^r$ (this follows from (2.44)) we see that

$$s_j^r = \sum_{k=0}^{j} A_{j-k}^{r-1}(x_0 + \dots x_k) = \sum_{k=0}^{j} x_k (A_0^r + \dots A_{j-k}^r) = \sum_{k=0}^{j} A_{j-k}^r x_k . \tag{2.45}$$

In particular, we obtain from (2.43) and (2.45) that

$$\frac{1}{(1-z)^{r+1}} \sum_{n=0}^{\infty} x_n z^n = \sum_{n=0}^{\infty} \left(\sum_{k=0}^{n} A_{n-k}^r x_k \right) z^n = \sum_{n=0}^{\infty} s_n^r z^n ; \tag{2.46}$$

therefore

$$\sum_{n=0}^{\infty} s_n^{r_1+r_2+1} z^n = \left(\frac{1}{(1-z)^{r_1+1}} \sum_{n=0}^{\infty} x_n z^n \right) \frac{1}{(1-z)^{r_2+1}}$$

$$= \sum_{n=0}^{\infty} s_n^{r_1} z^n \sum_{n=0}^{\infty} A_n^{r_2} z^n = \sum_{n=0}^{\infty} \left(\sum_{k=0}^{n} s_k^{r_1} A_{n-k}^{r_2} \right) z^n ,$$

implying the identities

$$s_n^{r_1+r_2+1} = \sum_{k=0}^{n} A_{n-k}^{r_2} s_k^{r_1} . \tag{2.47}$$

Furthermore, for $r \neq -1, -2, \dots$ an easy computation shows the following well-known equality

$$A_n^r = \frac{n^r}{\Gamma(r+1)} (1 + o(1)) . \tag{2.48}$$

After this preparation we are able to improve Theorem 9 for Cesàro summation of arbitrary order $r > 0$.

Theorem 10. *Let* $r > 0$, *and* $1 \le p < \infty, 1 \le q \le \infty$. *Then the matrix* $A = C^r \Sigma D_{(1/\log\log k)}$ *given by*

$$
a_{jk} = \begin{cases} \dfrac{A^r_{j-k}}{A^r_j} \dfrac{1}{\log\log k} & k \le j \\ 0 & k > j \end{cases}
$$

is (p,q)-*maximizing. For* $q < p$ *the* log-*term is superfluous.*

Note again that the last statement on the log-term is a special case of (the last statement in) Theorem 7.

Let us prove Theorem 10. As in the preceding section (see Theorem 1 and Theorem 5) we only have to show that there is some constant $C > 0$ such that for each orthonormal system (x_k) in some $L_2(\mu)$ and each sequence (α_k) of scalars we have

$$
\left\| \sup_j \left| \sum_{k=0}^{j} \frac{A^{r-1}_{j-k}}{A^r_j} \sum_{\ell=0}^{k} \alpha_\ell x_\ell \right| \right\|_2 \le C \| (\alpha_k \log\log k) \|_2 . \tag{2.49}
$$

Fix such (x_k) and (α_k), and recall from (2.45) that in our special situation

$$
s^r_j = \sum_{k=0}^{j} A^{r-1}_{j-k} \sum_{\ell=0}^{k} \alpha_\ell x_\ell = \sum_{k=0}^{j} A^r_{j-k} \alpha_k x_k , \tag{2.50}
$$

and

$$
\sigma^r_j = \frac{1}{A^r_j} s^r_j .
$$

By Theorem 9 the case $r = 1$ in (2.49) is already proved, and the case $r > 1$ is an immediate consequence of the next lemma (see also [1]).

Lemma 9. *Let* $r > -1$ *and* $\varepsilon > 0$. *Then*

$$
\left\| \sup_j | \sigma^{r+\varepsilon}_j | \right\|_2 \le \left\| \sup_j | \sigma^r_j | \right\|_2 .
$$

Proof. From (2.47) we deduce that $s^{r+\varepsilon}_j = \sum_{k=0}^{j} A^{\varepsilon-1}_{j-k} s^r_k$, and from (2.44) that

$$
\frac{1}{A^{r+\varepsilon}_j} \sum_{k=0}^{j} A^r_k A^{\varepsilon-1}_{j-k} = 1 .
$$

Hence we conclude

$$\left|\sigma_j^{r+\varepsilon}\right| = \left|\sum_{k=0}^{j} \frac{A_k^r A_{j-k}^{\varepsilon-1}}{A_j^{r+\varepsilon}} \sigma_k^r\right| \le \sup_{0 \le k \le j} |\sigma_k^r|,$$

which clearly proves our claim.   □

The proof of (2.49) for $1 > r > 0$ is slightly more complicated, and will follow from two Tauberian type results (we analyze proofs from [1, p.77,110]).

Lemma 10.

(1) For $r > -1/2$ and $\varepsilon > 0$

$$\left\|\sup_j |\sigma_j^{r+\frac{1}{2}+\varepsilon}|\right\|_2^2 \le C \left\|\sup_j \frac{1}{j+1} \sum_{k=0}^{j} |\sigma_k^r|^2\right\|_1.$$

(2) For $r > 1/2$

$$\left\|\sup_j \frac{1}{j+1} \sum_{k=0}^{j} |\sigma_k^{r-1}|^2\right\|_1 \le C \left(\|\alpha\|_2^2 + \left\|\sup_j |\sigma_j^r|\right\|_2^2\right).$$

Proof. For (1) note that by (2.47) and the Cauchy-Schwarz inequality

$$\left|\sigma_j^{r+\frac{1}{2}+\varepsilon}\right|^2 \le \sum_{k=0}^{j} |\sigma_k^r|^2 \frac{1}{(A_j^{r+\frac{1}{2}+\varepsilon})^2} \sum_{k=0}^{j} (A_k^r A_{j-k}^{-\frac{1}{2}+\varepsilon})^2,$$

and by (2.48) (for $j \ge 1$)

$$\frac{1}{(A_j^{r+\frac{1}{2}+\varepsilon})^2} \sum_{k=0}^{j} (A_k^r A_{j-k}^{-\frac{1}{2}+\varepsilon})^2 \le C_1 \frac{j^{2r}}{j^{2r+1+2\varepsilon}} \sum_{k=0}^{j} k^{-1+2\varepsilon} \le C \frac{1}{j+1},$$

the conclusion. For the proof of (2) note first that

$$\frac{1}{j+1} \sum_{k=0}^{j} |\sigma_k^{r-1}|^2 \le 2 \left(\frac{1}{j+1} \sum_{k=0}^{j} |\sigma_k^{r-1} - \sigma_k^r|^2 + \frac{1}{j+1} \sum_{k=0}^{j} |\sigma_k^r|^2\right),$$

hence for

$$\delta_j^r := \frac{1}{j+1} \sum_{k=0}^{j} |\sigma_k^{r-1} - \sigma_k^r|^2$$

we get that

$$\left\|\sup_j \frac{1}{j+1} \sum_{k=0}^{j} |\sigma_k^{r-1}|^2\right\|_1 \le 2 \left(\left\|\sup_j \delta_j^r\right\|_1 + \left\|\sup_j \frac{1}{j+1} \sum_{k=0}^{j} |\sigma_k^r|^2\right\|_1\right)$$

$$\le 2 \left(\left\|\sup_j \delta_j^r\right\|_1 + \left\|\sup_j |\sigma_j^r|\right\|_2^2\right).$$

It remains to check that $\left\| \sup_j \delta_j^r \right\|_1 \le C \|\alpha\|_2^2$. Since $A_n^r = A_n^{r-1} \frac{r+n}{r}$ we have by (2.50) that

$$
\sigma_j^r - \sigma_j^{r-1} = \sum_{k=0}^{j} \left(\frac{A_{j-k}^r}{A_j^r} - \frac{A_{j-k}^{r-1}}{A_j^{r-1}} \right) \alpha_k x_k
$$

$$
= \frac{1}{A_j^r A_j^{r-1}} \sum_{k=0}^{j} \left(A_{j-k}^r A_j^{r-1} - A_{j-k}^{r-1} A_j^r \right) \alpha_k x_k
$$

$$
= -\frac{1}{A_j^r} \sum_{k=0}^{j} \frac{k}{r} A_{j-k}^{r-1} \alpha_k x_k,
$$

hence by orthogonality

$$
\left\| \delta_{2^n}^r \right\|_1 = \frac{1}{2^n + 1} \sum_{j=0}^{2^n} \frac{1}{(A_j^r)^2} \sum_{k=0}^{j} \frac{k^2}{r^2} (A_{j-k}^{r-1})^2 |\alpha_k|^2
$$

$$
= \frac{1}{2^n + 1} \frac{1}{r^2} \sum_{k=0}^{2^n} k^2 |\alpha_k|^2 \sum_{j=k}^{2^n} \left(\frac{A_{j-k}^{r-1}}{A_j^r} \right)^2.
$$

From (2.48) we get

$$
\sum_{j=k}^{\infty} \left(\frac{A_{j-k}^{r-1}}{A_j^r} \right)^2 \le C_1 \sum_{j=k}^{\infty} \frac{(j-k)^{2r-2}}{j^{2r}}
$$

$$
\le C_1 \frac{1}{k^{2r}} \sum_{j=k}^{2k} (j-k)^{2r-2} + C_2 \sum_{j=2k+1}^{\infty} \frac{j^{2r-2}}{j^{2r}} \le C_3 \frac{1}{k}.
$$

But then

$$
\sum_{n=0}^{\infty} \left\| \delta_{2^n}^r \right\|_1 \le C_4 \sum_{n=0}^{\infty} \frac{1}{2^n + 1} \sum_{k=0}^{2^n} k |\alpha_k|^2
$$

$$
\le C_4 \sum_{k=0}^{\infty} k |\alpha_k|^2 \sum_{n:2^n \ge k} \frac{1}{2^{n+1}} \le C_5 \sum_{k=0}^{\infty} |\alpha_k|^2,
$$

which gives

$$
\left\| \sup_n \delta_n^r \right\|_1 \le 2 \left\| \sup_n \delta_{2^n}^r \right\|_1 \le C \sum_{k=0}^{\infty} |\alpha_k|^2.
$$

This completes the proof of (2). $\qquad\square$

Finally, we complete the

Proof (of (2.49) for $0 < r < 1$). By Theorem 9

$$\left\| \sup_j \sigma_j^1 \right\|_2 \le C \| (\alpha_k \log\log k) \|_2,$$

hence we deduce from Lemma 10 that for all $\varepsilon > 0$

$$\left\| \sup_j \sigma_j^{\frac{1}{2}+\varepsilon} \right\|_2^2 \le C_1 \left\| \sup_j \frac{1}{j+1} \sum_{k=0}^{j} |\sigma_k^0|^2 \right\|_1$$

$$\le C_2 \left(\|\alpha\|_2^2 + \left\| \sup_j \sigma_j^1 \right\|_2^2 \right) \le C_3 \| (\alpha_k \log\log k) \|_2^2.$$

A repetition of this argument gives

$$\left\| \sup_j \sigma_j^{2\varepsilon} \right\|_2^2 \le C_1 \left\| \sup_j \frac{1}{j+1} \sum_{k=0}^{j} |\sigma_k^{-\frac{1}{2}+\varepsilon}|^2 \right\|_1$$

$$\le C_2 \left(\|\alpha\|_2^2 + \left\| \sup_j \sigma_j^{\frac{1}{2}+\varepsilon} \right\|_2^2 \right) \le C_3 \| (\alpha_k \log\log k) \|_2^2,$$

the desired inequality.   □

This finishes the proof of Theorem 10, a result which in the form presented here is new – but let us mention again that the inequality (2.49) on orthonormal series behind Theorem 10 corresponds to the fundamental coefficient tests for Cesàro summation proved by Kaczmarz [43] and Menchoff [61, 62] (see also (1.7) and (1.8)). As in the Corollaries 1 and 3 we take advantage to add another natural scale of summing operators.

Corollary 4. *For $r > 0$ all matrices $A = C^r \Sigma D_{(1/\log\log k)}$ from Theorem 10, if considered as operators from ℓ_1 into ℓ_∞, are 1-summing.*

2.2.4 Kronecker Matrices

We now generate some matrices which later lead to laws of large numbers. The second part of the following simple lemma is usually known as Kronecker's lemma.

Lemma 11. *Let $A = (a_{jk})$ be a lower triangular matrix with entries in a Banach space X. Then*

(1) $\sum_{k=0}^{j} \frac{k}{j+1} a_{jk} = \sum_{k=0}^{j} a_{jk} - \frac{1}{j+1} \sum_{k=0}^{j} \sum_{\ell=0}^{k} a_{j\ell}$ for every j

(2) $\lim_j \frac{1}{j+1} \sum_{k=0}^{j} k a_{jk} = 0$ whenever $\left(\sum_{k=0}^{j} a_{jk} \right)_j$ converges

(3) *Let A be a lower triangle scalar matrix which is (p,q)-maximizing. Then the matrix B defined by*

$$b_{jk} := \begin{cases} \dfrac{k}{j+1} a_{jk} & k \le j \\ 0 & k > j \end{cases}$$

is again (p,q)-maximizing.

Proof. Statement (1) is immediate, and implies (2). In order to prove (3) apply (1) to see that for every choice of finitely many scalars $\xi_0, \ldots, \xi_j$ we have

$$\sup_j \left| \sum_{k=0}^{j} \frac{k}{j+1} a_{jk} \xi_k \right| \le 2 \sup_j \left| \sum_{k=0}^{j} a_{jk} \xi_k \right|,$$

and therefore by definition

$$m_{p,q}(B) \le 2 m_{p,q}(A),$$

the conclusion.   □

It makes sense to call matrices (b_{jk}) like in statement (3) Kronecker matrices – to see a first example, note that by Theorem 7 and the preceding lemma for any lower triangular summation process S the matrix

$$\left(\frac{k}{j+1} \frac{1}{\log k} \sum_{\ell=k}^{\infty} s_{j\ell} \right)_{j,k} \tag{2.51}$$

is (p,q)-maximizing. Sometimes the log-term can be improved – for example, for Cesàro summation of order $r > 0$; here we conclude from Theorem 10 that $\log k$ may be replaced by $\log \log k$. But the following theorem shows that in this case in fact no log-term at all is needed.

Theorem 11. *Let $1 \le p < \infty, 1 \le q \le \infty$. The matrix M defined by*

$$m_{jk} := \begin{cases} \dfrac{k}{j+1} \left(1 - \dfrac{k}{j+1}\right) & k \le j \\ 0 & k > j \end{cases} \tag{2.52}$$

is (p,q)-maximizing. More generally, for $r > 0$ the matrix M^r defined by

$$m_{jk}^r := \begin{cases} \dfrac{k}{j+1} \dfrac{A_{j-k}^r}{A_j^r} & k \le j \\ 0 & k > j \end{cases} \tag{2.53}$$

is (p,q)-maximizing.

Let us start with the proof of (2.52). Again we follow our general philosophy – we only show a maximal inequality for orthonormal series: Fix such a series $\sum_k \alpha_k x_k$ in $L_2(\mu)$, and put

$$\mu_j^0 = \sum_{k=0}^{j} \frac{k}{j+1} \alpha_k x_k \text{ and } \mu_j^1 = \sum_{k=0}^{j} \frac{k}{j+1}\left(1 - \frac{k}{j+1}\right)\alpha_k x_k.$$

In order to prove that M is (p,q)-maximizing, by Theorem 1 and Theorem 5 it suffices to show that

$$\left\| \sup_j |\mu_j^1| \right\|_2 \leq C \|\alpha\|_2, \tag{2.54}$$

$C > 0$ some universal constant. The proof of this inequality follows from a careful analysis of Moricz [63, Theorem 1]; similar to the proof of (2.28) and (2.38) we check three estimates:

$$\sum_{n=0}^{\infty} \| \mu_{2^n}^0 - \mu_{2^n}^1 \|_2^2 \leq C_1 \|\alpha\|_2^2 \tag{2.55}$$

$$\sum_{n=0}^{\infty} \left\| \max_{2^n < \ell \leq 2^{n+1}} |\mu_\ell^1 - \mu_{2^n}^1| \right\|_2^2 \leq C_2 \|\alpha\|_2^2 \tag{2.56}$$

$$\sum_{n=0}^{\infty} \| \mu_{2^n}^0 \|_2^2 \leq C_3 \|\alpha\|_2^2 ; \tag{2.57}$$

indeed, for $2^m < j \leq 2^{m+1}$

$$|\mu_j^1|^2 \leq \left(|\mu_j^1 - \mu_{2^m}^1| + |\mu_{2^m}^0 - \mu_{2^m}^1| + |\mu_{2^m}^0| \right)^2$$

$$\leq 3 \left(|\mu_j^1 - \mu_{2^m}^1|^2 + |\mu_{2^m}^0 - \mu_{2^m}^1|^2 + |\mu_{2^m}^0|^2 \right)$$

$$\leq 3 \left(\sum_{n=0}^{\infty} \max_{2^n < \ell \leq 2^{n+1}} |\mu_\ell^1 - \mu_{2^n}^1|^2 + \sum_{n=0}^{\infty} |\mu_{2^n}^0 - \mu_{2^n}^1|^2 + \sum_{n=0}^{\infty} |\mu_{2^n}^0|^2 \right).$$

Since the right side is independent of j, we obtain

$$\left\| \sup_j |\mu_j^1| \right\|_2^2 \leq 3 \left(\sum_{n=0}^{\infty} \left\| \max_{2^n < \ell \leq 2^{n+1}} |\mu_j^1 - \mu_{2^n}^1| \right\|_2^2 + \sum_{n=0}^{\infty} \| \mu_{2^n}^0 - \mu_{2^n}^1 \|_2^2 + \sum_{n=0}^{\infty} \| \mu_{2^n}^0 \|_2^2 \right),$$

which by (2.55), (2.56), and (2.57) gives the conclusion. (2.55) follows by orthogonality from

$$\sum_{n=0}^{\infty} \| \mu_{2^n}^0 - \mu_{2^n}^1 \|_2^2 = \sum_{n=0}^{\infty} \int \left| \frac{1}{(2^n+1)^2} \sum_{k=0}^{2^n} k^2 \alpha_k x_k \right|^2 d\mu$$

$$= \sum_{n=0}^{\infty} \frac{1}{(2^n+1)^4} \sum_{k=0}^{2^n} k^4 |\alpha_k|^2$$

$$\leq \sum_{k=0}^{\infty} |\alpha_k|^2 \sum_{n:2^n \geq k} \left(\frac{k}{2^n}\right)^4$$

$$= \sum_{k=0}^{\infty} |\alpha_k|^2 \sum_{n \geq \log k} \left(\frac{1}{n - 2^{\log k}}\right)^4 \leq C_1 \|\alpha\|_2^2,$$

and (2.57) from

$$\sum_{n=0}^{\infty} \int \left|\frac{1}{2^n+1} \sum_{k=0}^{2^n} k \alpha_k x_k\right|^2 d\mu = \sum_{n=0}^{\infty} \frac{1}{(2^n+1)^2} \sum_{k=0}^{2^n} k^2 |\alpha_k|^2$$

$$\leq \sum_{k=0}^{\infty} |\alpha_k|^2 \sum_{n:2^n \geq k} \left(\frac{k}{2^n}\right)^2 \leq C_3 \|\alpha\|_2^2.$$

For the proof of (2.56) note that for $2^n < \ell \leq 2^{n+1}$ by the Cauchy-Schwarz inequality

$$|\mu_\ell^1 - \mu_{2^n}^1| \leq \sum_{j=2^n+1}^{2^{n+1}} 1 \cdot |\mu_j^1 - \mu_{j-1}^1| \leq 2^{n/2} \left(\sum_{j=2^n+1}^{2^{n+1}} |\mu_j^1 - \mu_{j-1}^1|^2\right)^{\frac{1}{2}},$$

and hence

$$\max_{2^n < \ell \leq 2^{n+1}} |\mu_\ell^1 - \mu_{2^n}^1|^2 \leq 2^n \sum_{j=2^n+1}^{2^{n+1}} |\mu_j^1 - \mu_{j-1}^1|^2.$$

Since

$$\mu_j^1 - \mu_{j-1}^1 = \sum_{k=0}^{j} \frac{k}{j+1} \left(1 - \frac{k}{j+1}\right) \alpha_k x_k - \sum_{k=0}^{j-1} \frac{k}{j} \left(1 - \frac{k}{j}\right) \alpha_k x_k$$

$$= \sum_{k=0}^{j} \frac{k j^2 (j+1) - k^2 j^2 - k j (j+1)^2 + k^2 (j+1)^2}{j^2 (j+1)^2} \alpha_k x_k$$

$$= \sum_{k=0}^{j} \left(\frac{k^2 (2j+1)}{j^2 (j+1)^2} - \frac{k}{j(j+1)}\right) \alpha_k x_k$$

and for $k \leq j$

$$\frac{k^2 (2j+1)}{j^2 (j+1)^2} \leq \frac{k(2j+1)}{j(j+1)^2} \leq \frac{2k}{j(j+1)},$$

we have

$$\int |\mu_j^1 - \mu_{j-1}^1|^2 d\mu \leq \int \left|\sum_{k=0}^{j} \frac{k}{j(j+1)} \alpha_k x_k\right|^2 d\mu = \sum_{k=0}^{j} \left(\frac{k}{j(j+1)}\right)^2 |\alpha_k|^2.$$

Hence

$$\int \max_{2^n < \ell \le 2^{n+1}} |\mu_\ell^1 - \mu_{2^n}^1|^2 d\mu \le 2^n \sum_{j=2^n+1}^{2^{n+1}} \sum_{k=0}^{j} \frac{k^2}{j^2(j+1)^2} |\alpha_k|^2$$

$$\le \frac{1}{(2^n+1)^2} \sum_{j=2^n+1}^{2^{n+1}} \sum_{k=0}^{j} \frac{k^2}{j} |\alpha_k|^2$$

$$\le \frac{1}{(2^n+1)^2} \sum_{k=0}^{2^{n+1}} k^2 |\alpha_k|^2 \sum_{j=\max(k,2^n+1)}^{2^{n+1}} \frac{1}{j}$$

$$\le \frac{1}{(2^n+1)^2} \sum_{k=0}^{2^{n+1}} k^2 |\alpha_k|^2 \sum_{j=2^n+1}^{2^{n+1}} \frac{1}{j}$$

$$\le \frac{1}{(2^n+1)^2} \sum_{k=0}^{2^{n+1}} k^2 |\alpha_k|^2 .$$

This finishes the proof of (2.56):

$$\sum_{n=0}^{\infty} \| \max_{2^n < \ell \le 2^{n+1}} |\mu_\ell^1 - \mu_{2^n}^1| \|_2^2 \le \sum_{n=0}^{\infty} \frac{1}{(2^n+1)^2} \sum_{k=0}^{2^{n+1}} k^2 |\alpha_k|^2$$

$$\le \sum_{k=0}^{\infty} |\alpha_k|^2 \sum_{n:2^n \ge k} \left(\frac{k}{2^n}\right)^2 \le C_2 \|\alpha\|_2^2 ,$$

completing the proof of (2.52).

In order to prove (2.53) we follow the method from Sect. 2.2.3. Once again, we fix an orthonormal series $\sum_k \alpha_k x_k$ in $L_2(\mu)$ and show, according to Theorem 1 and Theorem 5, that for

$$s_j^r = \sum_{k=0}^{j} A_{j-k}^{r-1} \sum_{\ell=0}^{k} \ell \, \alpha_\ell x_\ell = \sum_{k=0}^{j} A_{j-k}^r k \, \alpha_k x_k \tag{2.58}$$

(for this last inequality see (2.45)) and

$$\mu_j^r = \frac{1}{(j+1)A_j^r} s_j^r ,$$

the following maximal inequality holds:

$$\| \sup_j |\mu_j^r| \|_2 \le C \|\alpha\|_2 . \tag{2.59}$$

Again the proof follows from an analysis of the work of Moricz in [63, Theorem 2]. In fact, the proof is very similar to the one of Theorem 10 (note the similarity of

(2.58) with (2.50)) – for the sake of completeness we give some details. With (2.52) the case $r = 1$ is again already proved, and the case $r > 0$ follows from two analogs of Lemma 9 and Lemma 10.

Lemma 12. *Let $r > -1$ and $\varepsilon > 0$. Then*

$$\left\| \sup_j |\mu_j^{r+\varepsilon}| \right\|_2 \leq \left\| \sup_j |\mu_j^r| \right\|_2 .$$

Proof. By (2.47) (as in the proof of Lemma 9) we have

$$\mu_j^{r+\varepsilon} = \frac{1}{(j+1)A_j^{r+\varepsilon}} \sum_{k=0}^{j} s_k^r A_{j-k}^{\varepsilon-1}$$

$$= \frac{1}{(j+1)A_j^{r+\varepsilon}} \sum_{k=0}^{j} \mu_k^r (k+1) A_k^r A_{j-k}^{\varepsilon-1} = \sum_{k=0}^{j} \beta_{jk} \mu_k^r$$

with $\beta_{jk} = \frac{(k+1)A_k^r A_{j-k}^{\varepsilon-1}}{(j+1)A_j^{r+\varepsilon}} \geq 0$, and for these coefficients (use again (2.44))

$$\sum_{k=0}^{j} \beta_{jk} \leq \frac{1}{A_j^{r+\varepsilon}} \sum_{k=0}^{j} A_k^r A_{j-k}^{\varepsilon-1} = 1 .$$

Hence the conclusion is immediate.   □

Lemma 13.

(1) For $r > -1/2$ and $\varepsilon > 0$

$$\left\| \sup_j |\mu_j^{r+\frac{1}{2}+\varepsilon}| \right\|_2^2 \leq C \left\| \sup_j \frac{1}{j+1} \sum_{k=0}^{j} |\mu_k^r|^2 \right\|_1 .$$

(2) For $r > 1/2$

$$\left\| \sup_j \frac{1}{j+1} \sum_{k=0}^{j} |\mu_k^{r-1}|^2 \right\|_1 \leq C \left(\|\alpha\|_2^2 + \left\| \sup_j |\mu_j^r| \right\|_2^2 \right) .$$

Proof. Observe first (see the preceding proof) that

$$\mu_j^{r+\frac{1}{2}+\varepsilon} = \frac{1}{(j+1)A_j^{r+\frac{1}{2}+\varepsilon}} \sum_{k=0}^{j} \mu_k^r (k+1) A_k^r A_{j-k}^{-\frac{1}{2}+\varepsilon} ,$$

hence as in the proof of Lemma 10 (by the Cauchy-Schwarz inequality) we get

$$|\mu_j^{r+\frac{1}{2}+\varepsilon}|^2 \le C \frac{1}{j+1} \sum_{k=0}^{j} |\mu_k^r|^2 .$$

For the proof of (2) define

$$\delta_j^r := \frac{1}{j+1} \sum_{k=0}^{j} |\mu_k^{r-1} - \mu_k^r|^2 ,$$

and show as in the proof of Lemma 10 first

$$\left\| \sup_j \frac{1}{j+1} \sum_{k=0}^{j} |\mu_k^{r-1}|^2 \right\|_1 \le 2 \left(\left\| \sup_j \delta_j^r \right\|_1 + \left\| \sup_j |\mu_j^r|^2 \right\|_2^2 \right),$$

and then

$$\sum_{n=0}^{\infty} \left\| \delta_{2^n}^r \right\|_1 \le C_1 \sum_{k=0}^{\infty} \frac{1}{2^n+1} \sum_{k=0}^{2^n} k |\alpha_k|^2$$

$$\le C_1 \sum_{k=0}^{\infty} k |\alpha_k|^2 \sum_{n:2^n \ge k} \frac{1}{2^{n+1}} \le C_2 \sum_{k=0}^{\infty} |\alpha_k|^2 ,$$

which again implies the conclusion easily. $\square$

Finally, we deduce (2.59) (and hence complete the proof of Theorem 11) word by word as this was done in the proof of Theorem 10 (or better (2.49)) at the end of Sect. 2.2.3.

2.2.5   Abel Matrices

The following result on Abel matrices A^ρ (see Sect. 2.1.1 for the definition) is a straight forward consequence of our results on Cesàro summation.

Theorem 12. *Let (ρ_j) be a positive and strictly increasing sequence converging to 1. Then the matrix $A = A^\rho \Sigma D_{(1/\log\log k)}$ given by*

$$a_{jk} = \frac{\rho_j^k}{\log\log k}$$

is (p,q)-maximizing. Again for $q < p$ no log-term is needed.

Proof. The proof is standard, we rewrite the matrix A in terms of the Cesàro matrix. We have for all j and every choice of finitely many scalars $x_0, \ldots, x_n$ that (use (2.46) for $r = 1$)

$$\sum_{k=0}^{n} \rho_j^k x_k = (1 - \rho_j)^2 \sum_{k=0}^{n} s_k^1 \rho_j^k$$

$$= (1 - \rho_j)^2 \sum_{k=0}^{n} \rho_j^k \sum_{\ell=0}^{k} s_\ell$$

$$= \sum_{k=0}^{n} (1 - \rho_j)^2 \rho_j^k (k+1) \frac{1}{k+1} \sum_{\ell=0}^{k} \sum_{m=0}^{\ell} x_m .$$

Define the matrix S through $s_{jk} = (1 - \rho_j)^2 \rho_j^k (k+1)$. By (2.43) we know that $\sum_k s_{jk} = 1$ so that S defines a bounded operator on ℓ_∞. Since we just proved that $A^\rho = SC$, the conclusion now follows from Theorem 10 (compare the maximal functions as in (2.11)). The last statement is a special case of Theorem 2. □

As a sort of by product we obtain from Theorem 4 a further interesting scale of 1-summing operators from ℓ_1 to ℓ_∞ (see also the Corollaries 1, 3, and 4).

Corollary 5. *All matrices $A = A^\rho \Sigma D_{(1/\log\log k)}$ form 1-summing operators from ℓ_1 into ℓ_∞.*

2.2.6 Schur Multipliers

We sketch without any proofs that our setting of maximizing matrices is equivalent to Bennett's theory of (p, q)-Schur multipliers from [3]; for precise references see the notes and remarks at the end of this section. As mentioned above our theory of maximizing matrices was up to some part modeled along this theory.

An infinite matrix $M = (m_{jk})_{j,k \in \mathbb{N}_0}$ with $\|M\|_\infty < \infty$ is said to be a (p, q)-multiplier $(1 \leq p, q \leq \infty)$ if its Schur product $M * A = (m_{jk} a_{jk})_{j,k}$ with any infinite matrix $A = (a_{jk})_{j,k \in \mathbb{N}_0}$ maps ℓ_p into ℓ_q whenever A does. In this case, the (p, q)-multiplier norm of M is defined to be

$$\mu_{p,q}(M) = \sup \|M * A : \ell_p \to \ell_q\| ,$$

the infimum taken over all matrices A which define operators from ℓ_p into ℓ_q of norm ≤ 1. For $p = q = 2$ we simply speak of multipliers; we remark that

$$\mu_{2,2}(M) = \|M\|_{cb} , \tag{2.60}$$

where $\|M\|_{cb}$ denotes the completely bounded norm of M which via Schur multiplication is considered as an operator on the operator space $\mathscr{L}(\ell_2)$.

Moreover, it is known that the (p,q)-multiplier norm has the following formulation in terms of summing norms:

$$\mu_{p,q}(M) = \sup_{\|\alpha\|_{\ell_p} \leq 1} \pi_q(M D_\alpha) \tag{2.61}$$

(here M is considered as an operator from ℓ_1 into ℓ_∞, and $D_\alpha : \ell_{p'} \to \ell_1$ denotes again the diagonal operator associated to α). From Theorem 3 we conclude that M is a (p,q)-Schur multplier if and only if M is (q,p)-maximizing – with equal norms:

$$\mu_{p,q}(M) = m_{q,p}(M). \tag{2.62}$$

This in particular means that all facts of the rich theory of Schur multipliers apply to maximizing operators, and vice versa. We mention some consequences, of course focusing on maximizing matrices:

(1) Obviously, $\mu_{p,q}(M) = \mu_{q',p'}(M^t)$, where M^t is the transposed (or dual) matrix of M, hence by (2.62) we have

$$m_{p,q}(M) = m_{q',p'}(M^t).$$

By definition it is obvious that (p,q)-maximizing matrices are insensitive with respect to row repetitions or row permutations, i.e. if A is (p,q)-maximizing, then

$$m_{p,q}(A) = m_{p,q}(\tilde{A})$$

where $\tilde{A}$ is obtained from A by repeating or permuting rows. By transposition, we see that $m_{p,q}$ is insensitive to column repetitions or permutations.

(2) For two (p,q)-maximizing matrices A and B their Schur product $A * B$ is again (p,q)-maximizing, and

$$m_{p,q}(A * B) = m_{p,q}(A)\, m_{p,q}(B)$$

(a fact obvious for Schur multipliers). A similar result holds for tensor products (Kronecker products) of Schur multipliers,

$$m_{p,q}(A \otimes B) \leq m_{p,q}(A)\, m_{p,q}(B).$$

(3) Denote by T_n the nth-main triangle projection, i.e. the projection on the vector space of all infinite matrices $A = (a_{jk})_{j,k \in \mathbb{N}_0}$ with $\|A\|_\infty < \infty$ defined by

$$T_n(A) := \sum_{j+k \leq n} a_{jk} e_j \otimes e_k;$$

obviously, $T_n(A) = A * \Theta_n$, where

$$\Theta_n(j,k) := \begin{cases} 1 & j+k \leq n \\ 0 & \text{elsewhere}. \end{cases}$$

Then we conclude from (2.62) that for arbitrary p,q

$$\mu_{q,p}(\Theta_n) = \mathrm{m}_{p,q}(\Theta_n) = \mathrm{m}_{p,q}(\Sigma_n),$$

where Σ_n again is the sum matrix (see (2.30)); here the last equality is obvious by the definition of the (p,q)-maximizing norm. From Theorem 2 and the estimate from (2.31) (use also (2.8)) we deduce that for some constant C independent of n

$$\mu_{q,p}(\Theta_n) \leq \begin{cases} C\log n & p \leq q \\ C & q < p. \end{cases}$$

(4) Recall that a matrix $M = (m_{jk})_{j,k \in \mathbb{N}_0}$ is said to be a Toeplitz matrix whenever it has the form $m_{jk} = c_{j-k}$ with $c = (c_n)_{n \in \mathbb{Z}}$ a scalar sequence. A Toeplitz matrix is $(2,2)$-maximizing if and only if there exists a bounded complex Borel measure μ on the torus $\mathbb{T}$ such that its Fourier transform $\hat{\mu}$ equals c.

(5) Denote by $\mathscr{C}$ the closed convex hull of the set of all $(2,2)$-maximizing matrices A of the form $a_{jk} = \alpha_j \beta_k$, where α and β are scalar sequences bounded by 1 and the closure is taken in the coordinatewise topology. Then we have that

$$\mathscr{C} \subset \{A \mid \mathrm{m}_{2,2}(A) \leq 1\} \subset K_G \mathscr{C},$$

K_G Grothendieck's constant.

Notes and remarks: The close connection of Schur multipliers and summing operators was observed and elaborated by many authors. See for example Grothendieck [21], Kwapień-Pełczyński [50] and, very important here, Bennett's seminal paper [3] which is full of relevant information for our purpose and motivated large part of this second chapter. Equation (2.61) is Bennett's Theorem [3, Sect. 4.3], and (2.60) is a result due to Haagerup [25]. For Schur multipliers instead of maximizing matrices the Theorems 4 (note that its analog from [3, Theorem 6.4] for multipliers is weaker and contains a wrong statement on the constant) and 5 are well-known; for $p = q$ see [78, Theorem 5.11] and the notes and remarks thereby (Pisier:" Once Kwapień had extended the factorization theorems to the L_p-case, it is probably fair to say that it was not too difficult to extend the theory of Schur mulipliers …"). Remark (1), (2) and (4) from Sect. 2.2.6 are taken from Bennett [3] (there of course formulated for Schur multipliers instead of maximizing martrices), and Remark (5) from [78]. For the final estimate in 2.2.6, (3) see [2] and [50].

2.3   Limit Theorems in Banach Function Spaces

It is remarkable that most of the classical almost everywhere summation theorems for orthogonal series in $L_2(\mu)$ without too many further assumptions in a natural way extend to vector-valued Banach function spaces $E(\mu, X)$.

We illustrate that our setting of maximizing matrices in a very comfortable way leads not only to the most important classical results, but also to strong new extensions of them. We show, as announced earlier, that most of the classical coefficient tests on pointwise summation of orthogonal series – in particular those for Cesàro, Riesz and Abel summation – together with their related maximal inequalities, have natural analogs for the summation of unconditionally convergent series in vector-valued Banach function spaces $E(\mu, X)$.

The main results are collected in the Theorems 13 and 14, and then later applied to classical summation methods (see the Corollaries 6 and 7). Moreover, we prove that each unconditionally convergent series in $L_p(\mu)$ is Riesz$^\lambda$-summable for some sequence λ; this is an L_p-analog of an important observation on orthonormal series apparently due to Alexits [1, p.142]. We finish this section with a systematic study of laws of large numbers in vector-valued Banach function spaces $E(\mu, X)$ with respect to arbitrary summation methods – in particular we extend some "non logarithmical" laws of large numbers due to Moricz [64].

2.3.1   Coefficient Tests in Banach Function Spaces

We start with a description of the situation in L_p-spaces – here the main step is a rather immediate consequence of our general frame of maximizing matrices:

Assume that S is a summation method and ω a Weyl sequence (see (2.2) for the defintion) such that for each orthonormal series $\sum_k \alpha_k x_k$ in $L_2(\mu)$ we have that the maximal function of the linear means

$$\sum_{k=0}^{\infty} s_{jk} \sum_{\ell=0}^{k} \frac{\alpha_\ell}{\omega_\ell} x_\ell, \; j \in \mathbb{N}_0$$

is square integrable,

$$\sup_j \left| \sum_{k=0}^{\infty} s_{jk} \sum_{\ell=0}^{k} \frac{\alpha_\ell}{\omega_\ell} x_\ell \right| \in L_2(\mu) \; ; \tag{2.63}$$

this implicitly means to assume that we are in one of the classical situations described above. How can this result be transferred to L_p-spaces, $1 \le p < \infty$?

By Theorem 1 our assumption means precisely that the matrix $A = S\Sigma D_{1/\omega}$ is $(2,2)$-maximizing. As a consequence A by Theorem 5 even is (p, ∞)-maximizing,

$1 \leq p < \infty$, i.e. for each unconditionally convergent series $\sum_k x_k$ in $L_p(\mu)$ we have that

$$\sup_j \left| \sum_{k=0}^{\infty} s_{jk} \sum_{\ell=0}^{k} \frac{x_\ell}{\omega_\ell} \right| \in L_p(\mu), \tag{2.64}$$

or equivalently in terms of an inequality, there is a constant $C > 0$ such that for each such series

$$\left\| \sup_j \left| \sum_{k=0}^{\infty} s_{jk} \sum_{\ell=0}^{k} \frac{x_\ell}{\omega_\ell} \right| \right\|_p \leq C w_1(x_k).$$

But then we deduce from Proposition 2 that for each unconditionally convergent series $\sum_k x_k$ in $L_p(\mu)$

$$\sum_{k=0}^{\infty} \frac{x_k}{\omega_k} = \lim_j \sum_{k=0}^{\infty} s_{jk} \sum_{\ell=0}^{k} \frac{x_\ell}{\omega_\ell} \quad \mu\text{-a.e.} \tag{2.65}$$

To summarize, if we start with a classical pointwise summation theorem on orthogonal series and know in addition that the underlying summation method even allows a maximal theorem for these series like in (2.63), then we obtain with (2.64) and (2.65) a strong extension of this result in L_p-spaces. Based on tensor products we now even prove that here $L_p(\mu)$ can be replaced by an arbitrary vector-valued Banach function space $E(\mu, X)$, and this without any further assumption on the function space $E(\mu)$ or Banach space X.

Theorem 13. *Let $E(\mu)$ be a Banach function space, X a Banach space, and $A = (a_{jk})$ a $(2,2)$-maximzing matrix. Then for each unconditionally convergent series $\sum_k x_k$ in $E(\mu, X)$ the following statements hold:*

(1) $\sup_j \left\| \sum_{k=0}^{\infty} a_{jk} x_k(\cdot) \right\|_X \in E(\mu)$

(2) The sequence $\left(\sum_{k=0}^{\infty} a_{jk} x_k \right)_j$ converges μ-a.e. provided (a_{jk}) converges in each column.

In particular, let S be a summation method and ω a Weyl sequence with the additional property that for each orthonormal series $\sum_k \alpha_k x_k$ in $L_2(\mu)$ we have

$$\sup_j \left| \sum_{k=0}^{\infty} s_{jk} \sum_{\ell=0}^{k} \frac{\alpha_\ell}{\omega_\ell} x_\ell \right| \in L_2(\mu).$$

Then for each unconditionally convergent series $\sum_k x_k$ in $E(\mu, X)$ the following two statements hold:

(3) $\sup_j \left\| \sum_{k=0}^{\infty} s_{jk} \sum_{\ell=0}^{k} \frac{x_\ell(\cdot)}{\omega_\ell} \right\|_X \in E(\mu)$

(4) $\sum_{k=0}^{\infty} \frac{x_k}{\omega_k} = \lim_j \sum_{k=0}^{\infty} s_{jk} \sum_{\ell=0}^{k} \frac{x_\ell}{\omega_\ell} \quad \mu$-a.e.

Proof. In order to establish (1) we prove that for all n

$$\| \mathrm{id}_{E(\mu,X)} \otimes A_n : E(\mu,X) \otimes_\varepsilon \ell_1^n \longrightarrow E(\mu,X)[\ell_\infty^n]\| \leq K_G \, \mathrm{m}_{2,2}(A) , \qquad (2.66)$$

where A_n equals A for all entries a_{jk} with $1 \leq j,k \leq n$ and is zero elsewhere, and K_G again stands for Grothendieck's constant. Indeed, this gives our conclusion: For a finite sequence $(x_k)_{k=0}^n \in E(\mu,X)^{n+1}$ we have

$$w_1(x_k) = \left\| \sum_{k=0}^n x_k \otimes e_k \right\|_{E(\mu,X)\otimes_\varepsilon \ell_1^n}$$

(direct calculation) and

$$\left(\mathrm{id}_{E(\mu,X)} \otimes A_n \right) \left(\sum_k x_k \otimes e_k \right) = \sum_k x_k \otimes A_n(e_k)$$

$$= \sum_k x_k \otimes \sum_j a_{jk}\alpha_k e_j$$

$$= \sum_j \left(\sum_k a_{jk}x_k \right) \otimes e_j ,$$

therefore

$$\left\| \mathrm{id}_{E(\mu,X)} \otimes A_n \left(\sum_k x_k \otimes e_k \right) \right\|_{E(\mu,X)[\ell_\infty^n]} = \left\| \sup_j \left\| \sum_k a_{jk}x_k(\cdot) \right\|_X \right\|_{E(\mu)} .$$

Hence we have shown that for every choice of scalars $\alpha_0, \ldots, \alpha_n$ and functions $x_0, \ldots, x_n \in E(\mu,X)$

$$\left\| \sup_j \left\| \sum_k a_{jk}x_k(\cdot) \right\|_X \right\|_{E(\mu)} \leq K_G \, \mathrm{m}_{2,2}(A) w_1(x_k) ,$$

which then by Lemma 3 allows to deduce the desired result on infinite sequences. In order to prove (2.66) note first that by (2.24) and again Theorem 4 we have

$$\iota(A_n) \leq K_G \, \gamma_2(A_n) = K_G \, \mathrm{m}_{2,2}(A_n) .$$

Hence we deduce from (2.23) and Lemma 3 that

$$\| \mathrm{id}_{E(\mu,X)} \otimes A_n : E(\mu,X) \otimes_\varepsilon \ell_1^n \longrightarrow E(\mu,X) \otimes_\pi \ell_\infty^n\| \leq \iota(A_n) \leq K_G \, \mathrm{m}_{2,2}(A) ,$$

but since

$$\| \mathrm{id} : E(\mu,X) \otimes_\pi \ell_\infty^n \to E(\mu,X)[\ell_\infty^n]\| \leq 1 ,$$

this gives the desired estimate (2.66) and completes the proof of (1). The proof of statement (2) is now a consequence of Proposition 2. For a slightly different argument which avoids Lemma 3 see the proof of Theorem 17. Finally, for the proof of (3) and (4) define the matrix

$$A = S\Sigma D_{1/\omega}, \quad a_{jk} := \frac{1}{\omega_k}\sum_{\ell=k}^{\infty} s_{j\ell},$$

and note that for all j

$$\sum_{k=0}^{\infty} a_{jk}x_k = \sum_{k=0}^{\infty} s_{jk}\sum_{\ell=0}^{k}\frac{x_\ell}{\omega_\ell}.$$

Then we conclude by the assumption on S and Theorem 1 that A is $(2,2)$-maximizing which allows to deduce (3) from (2). Since by Proposition 1 for all k

$$\lim_{j} a_{jk} = \lim_{j}\frac{1}{\omega_k}\sum_{\ell=k}^{\infty} s_{j\ell} = \lim_{j}\frac{1}{\omega_k}\Big(\sum_{\ell=0}^{\infty} s_{j\ell} - \sum_{\ell=0}^{k-1} s_{j\ell}\Big) = \frac{1}{\omega_k}, \qquad (2.67)$$

statement (4) is consequence of (2).   □

To illustrate the preceding result, we collect some concrete examples on summation of unconditionally convergent series in vector-valued Banach function spaces. Note that in order to start the method one has to find appropriate maximal inequalties, i.e. to make sure that the matrices $S\Sigma D_{1/\omega}$ are $(2,2)$-maximizing (Theorem 1). In the literature most coefficient tests for almost sure summation (with respect to a summation method S and a Weyl sequence ω) do not come jointly with a maximal inequality. As mentioned, the maximal inequality (2.28) joining the Menchoff-Rademacher Theorem 6 was discovered much later by Kantorovitch in [46]. We showed in the preceding Sect. 2.2 that in many concrete situations the needed maximal inequalities follow from a careful analysis of the corresponding coefficient tests; for pure summation $S = \mathrm{id}$ this is the Kantorovitch-Menchoff-Rademacher inequality (2.28) from Theorem 6, for the Riesz method R^λ see (2.38) inducing Theorem 8, for the Cesàro method of order r (2.49) inducing Theorem 10, and finally Theorem 12 for the Abel method.

Corollary 6. *Let $\sum_k x_k$ be an unconditionally convergent series in a vector-valued Banach function space $E(\mu,X)$. Then*

(1) $\sup_j\Big\|\sum_{k=0}^{j}\frac{x_k(\cdot)}{\log k}\Big\|_X \in E(\mu)$

(2) $\sup_j\Big\|\sum_{k=0}^{j}\frac{\lambda_{k+1}-\lambda_k}{\lambda_{j+1}}\sum_{\ell=0}^{k}\frac{x_\ell(\cdot)}{\log\log\lambda_\ell}\Big\|_X \in E(\mu)$ *for every strictly increasing,*
unbounded and positive sequence (λ_k) of scalars

(3) $\sup_j\Big\|\sum_{k=0}^{j}\frac{A_{j-k}^{r-1}}{A_j^r}\sum_{\ell=0}^{k}\frac{x_\ell(\cdot)}{\log\log\ell}\Big\|_X \in E(\mu)$ *for every $r>0$*

(4) $\sup_j\Big\|\sum_{k=0}^{\infty}\rho_j^k\frac{x_k(\cdot)}{\log\log k}\Big\|_X \in E(\mu)$ *for every positive strictly increasing*
sequence (ρ_j) converging to 1.

Moreover, in each of these cases

$$\sum_{k=0}^{\infty} \frac{x_k}{\omega_k} = \lim_{j} \sum_{k=0}^{\infty} s_{jk} \sum_{\ell=0}^{k} \frac{x_\ell}{\omega_\ell} \quad \mu-a.e.,$$

where the summation method S is either given by the identity, Riesz$^\lambda$, Cesàro^r, or Abel^p matrix, and ω is the related Weyl sequence from (1) *up to* (4).

For $E(\mu, X) = L_p(\mu)$ the origin of statement (1) in Corollary 6 lies in the article [50, Theorem 5.1] of Kwapień and Pełczyński where a slightly weaker result is shown. The final form of (1) in L_p-spaces is due to Bennett [2, Theorem 2.5, Corollary 2.6] and Maurey-Nahoum [59], and was reproved in [68]. Moreover, in this special situation, statement (2) is also due to Bennett [2, Theorem 6.4], whereas both statements (3) and (4) seem to be new. Recall that the underlying four classical coefficient tests for orthogonal series are well-known theorems by Kaczmarz [43], Kantorovitch [46], Menchoff [60, 61, 62], Rademacher [81], and Zygmund [97]. Finally, we mention that by use of Corollary 2 a "lacunary version" of statement (1) can be proved.

We now extend the preceding result considerably. A Banach function space $E(\mu)$ is said to be *p-convex* if there is some constant $C \geq 0$ such that for each choice of finitely many functions $x_1, \ldots, x_n \in E(\mu)$ we have

$$\left\| \left(\sum_k |x_k|^p \right)^{1/p} \right\|_{E(\mu)} \leq C \left(\sum_k \|x_k\|_{E(\mu)}^p \right)^{1/p}, \tag{2.68}$$

and the best such C is usually denoted by $M^{(p)}(E(\mu))$ (compare also with Sect. 3.1.1). We here only mention that every Banach space $L_p(\mu)$ is *p*-convex with constant 1, but there are numerous other examples as can be seen e.g. in [53, 54].

Theorem 14. *Let $A = (a_{jk})$ be a (p,q)-maximizing matrix, $E(\mu)$ a p-convex Banach function space, and X a Banach space. Then for every $\alpha \in \ell_q$ and every weakly q'-summable sequence (x_k) in $E(\mu, X)$ we have*

$$\sup_j \left\| \sum_{k=0}^{\infty} a_{jk} \alpha_k x_k(\cdot) \right\|_X \in E(\mu),$$

and moreover $\left(\sum_{k=0}^{\infty} a_{jk} \alpha_k x_k \right)_j$ converges μ-a.e. provided each column of A converges; for this latter statement assume that (x_k) is unconditionally summable whenever $q = \infty$.

Note that Theorem 14 still contains Theorem 13 as a special case: If A is $(2,2)$-maximizing, then we conclude from Theorem 5 that the matrix A is even $(1,\infty)$-maximizing. Since every Banach function space E is 1-convex, in this special situation no convexity condition is needed.

Proof. Again, it suffices to show that for every choice of finitely many scalars $\alpha_0, \ldots, \alpha_n$

$$\| \mathrm{id}_{E(\mu,X)} \otimes A_n D_\alpha : E(\mu,X) \otimes_\varepsilon \ell_{q'}^n \longrightarrow E(\mu,X)[\ell_\infty^n] \|$$

$$\leq M^{(p)}(E(\mu)) \, \mathrm{m}_{p,q}(A) \|\alpha\|_q \, , \tag{2.69}$$

where D_α stands for the induced diagonal operator, and A_n equals A for all entries a_{jk} with $0 \leq j,k \leq n$ and is zero elsewhere. Indeed, as above we then obtain the conclusion: For any finite sequence $(x_k)_{k=0}^n \in E(\mu,X)^{n+1}$ we have

$$w_{q'}(x_k) = \left\| \sum_{k=0}^n x_k \otimes e_k \right\|_{E(\mu,X) \otimes_\varepsilon \ell_{q'}^n}$$

and

$$\left(\mathrm{id}_{E(\mu,X)} \otimes A_n D_\alpha \right) \left(\sum_k x_k \otimes e_k \right) = \sum_j \left(\sum_k a_{jk} \alpha_k x_k \right) \otimes e_j .$$

Then

$$\left\| \mathrm{id}_{E(\mu,X)} \otimes A_n D_\alpha \left(\sum_k x_k \otimes e_k \right) \right\|_{E(\mu,X)[\ell_\infty^n]} = \left\| \sup_j \left\| \sum_k a_{jk} \alpha_k x_k(\cdot) \right\|_X \right\|_{E(\mu)},$$

and hence we obtain the inequality

$$\left\| \sup_j \left\| \sum_k a_{jk} \alpha_k x_k(\cdot) \right\|_X \right\|_{E(\mu)} \leq M^{(p)}(E(\mu)) \, \mathrm{m}_{p,q}(A) \, \|\alpha\|_q \, w_{q'}(x_k) .$$

Finally, this inequality combined with Lemma 3 gives the statement of the theorem. For the proof of (2.69) fix scalars $\alpha_0, \ldots, \alpha_n$. By the general characterization of (p,q)-maximizing matrices from Theorem 3 as well as (2.18) and (2.19), we obtain a factorization

$$
\begin{array}{ccc}
\ell_{q'}^n & \xrightarrow{\ A_n D_\alpha\ } & \ell_\infty^n \\
{\scriptstyle R}\downarrow & & \uparrow{\scriptstyle S} \\
\ell_\infty^m & \xrightarrow[\ D_\mu\ ]{} & \ell_p^m
\end{array}
$$

with

$$\|R\| \, \|D_\mu\| \, \|S\| \leq (1+\varepsilon) \, \iota_p(A D_\alpha) \leq (1+\varepsilon) \, \mathrm{m}_{p,q}(A) \|\alpha\|_q .$$

Tensorizing gives the commutative diagram

$$
\begin{array}{ccc}
E(\mu,X) \otimes_\varepsilon \ell_{q'}^n & \xrightarrow{\ \mathrm{id}_{E(\mu,X)} \otimes A_n D_\alpha\ } & E(\mu,X)[\ell_\infty^n] \\
\Big\downarrow{\scriptstyle \mathrm{id}_{E(\mu,X)} \otimes R} & & \Big\uparrow{\scriptstyle \mathrm{id}_{E(\mu,X)} \otimes S} \\
E(\mu,X) \otimes_\varepsilon \ell_\infty^m & \xrightarrow{\ \mathrm{id}_{E(\mu,X)} \otimes D_\mu\ } & \ell_p^m(E(\mu,X)) .
\end{array}
$$

By the metric mapping property of ε we have

$$
\left\| \mathrm{id}_{E(\mu,X)} \otimes R \right\| \le \|R\| ,
$$

and moreover

$$
\begin{aligned}
\left\| \mathrm{id}_{E(\mu,X)} \otimes D_\mu \left(\sum_{k=0}^m x_k \otimes e_k \right) \right\|_{\ell_p^m(E(\mu,X))} &= \left(\sum_{k=0}^m \| \mu_k x_k \|_{E(\mu,X)}^p \right)^{1/p} \\
&\le \sup_k \|x_k\|_{E(\mu,X)} \left(\sum_{k=0}^m |\mu_k|^p \right)^{1/p}
\end{aligned}
$$

implies

$$
\left\| \mathrm{id}_{E(\mu,X)} \otimes D_\mu \right\| \le \|D_\mu\| .
$$

We show that

$$
\left\| \mathrm{id}_{E(\mu,X)} \otimes S \right\| \le M^{(p)}(E(\mu)) \, \|S\| ; \tag{2.70}
$$

indeed, as an easy consequence of (2.68) and Hölder's inequality we obtain

$$
\begin{aligned}
\left\| \sup_{j=1,\ldots,n} \left\| \sum_{k=0}^m s_{jk} x_k(\cdot) \right\|_X \right\|_{E(\mu)} &\le \left\| \sup_{j=1,\ldots,n} \left(\sum_{k=0}^m |s_{jk}|^{p'} \right)^{1/p'} \left(\sum_{k=0}^m \|x_k(\cdot)\|_X^p \right)^{1/p} \right\|_{E(\mu)} \\
&= \sup_{j=1,\ldots,n} \left(\sum_{k=0}^m |s_{jk}|^{p'} \right)^{1/p'} \left\| \left(\sum_{k=0}^m \|x_k(\cdot)\|_X^p \right)^{1/p} \right\|_{E(\mu)} \\
&\le M^{(p)}(E(\mu)) \sup_{j=1,\ldots,n} \left(\sum_{k=0}^m |s_{jk}|^{p'} \right)^{1/p'} \left(\sum_{k=0}^m \|x_k\|_{E(\mu,X)}^p \right)^{1/p} ,
\end{aligned}
$$

and this completes the proof of (2.69). The result on almost everywhere convergence again follows by Proposition 2. $\qquad\square$

Corollary 6 presents analogs of classical coefficient theorems with logarithmic Weyl sequences for unconditionally convergent series in vector-valued Banach function spaces, e.g. on Cesàro or Riesz summation. The following result shows that under restrictions on the series and the underlying function space these logarithmic terms are superfluous.

Corollary 7. *Assume that* $1 \leq q < p < \infty$, *and let* $E(\mu)$ *be a p-convex Banach function space and* X *a Banach space. Then for each* $\alpha \in \ell_q$ *and each weakly* q'-*summable sequence* (x_k) *in* $E(\mu,X)$ *we have*

(1) $\sup_j \left\| \sum_{k=0}^{j} \alpha_k x_k(\cdot) \right\|_X \in E(\mu)$

(2) $\sup_j \left\| \sum_{k=0}^{j} \dfrac{\lambda_{k+1} - \lambda_k}{\lambda_{j+1}} \sum_{\ell=0}^{k} \alpha_\ell x_\ell(\cdot) \right\|_X \in E(\mu)$ *for every strictly increasing, unbounded and positive sequence* (λ_k) *of scalars*

(3) $\sup_j \left\| \sum_{k=0}^{j} \dfrac{A_{j-k}^{r-1}}{A_j^r} \sum_{\ell=0}^{k} \alpha_\ell x_\ell(\cdot) \right\|_X \in E(\mu)$ *for every* $r > 0$

(4) $\sup_j \left\| \sum_{k=0}^{\infty} \rho_j^k \alpha_\ell x_\ell(\cdot) \right\|_X \in E(\mu)$ *for every positive strictly increasing sequence* (ρ_j) *converging to* 1.

Moreover, in each of these cases

$$\sum_{k=0}^{\infty} \alpha_k x_k = \lim_j \sum_{k=0}^{\infty} s_{jk} \sum_{\ell=0}^{k} \alpha_\ell x_\ell(\cdot) \quad \mu - a.e.,$$

where the summation method S *is either given by the identity, Riesz$^\lambda$, Cesàro^r, or Abel$^\rho$ matrix.*

Proof. The argument by now is clear: For each of the considered summation methods the matrix $A = S\Sigma$ by Theorem 2 is (p,q)-maximizing. Since

$$\sum_{k=0}^{\infty} a_{jk} x_k = \sum_{k=0}^{\infty} s_{jk} \sum_{\ell=0}^{k} x_\ell,$$

Theorem 14 gives the conclusion. The result on μ-a.e. convergence again follows from Proposition 2.   □

Of course, the preceding corollary could also be formulated for arbitrary summation methods instead of the four concrete examples given here. Statement (1) is a far reaching extension of a well-known result of Menchoff [60] and Orlicz [67] for orthonormal series.

Finally, we present a sort of converse of Corollary 6,(2): The sum of every unconditionally convergent series $\sum_k x_k$ in $E(\mu,X)$ (such that $E(\mu)$ and X have finite cotype) can be obtained by almost everywhere summation of its partial sums through a properly chosen Riesz method. Recall that a Banach space X has cotype p, $2 \leq p < \infty$ whenever there is some constant $C \geq 0$ such that for each choice of finitely many vectors $x_1, \ldots, x_n \in X$ we have

$$\left(\sum_k \|x_k\|^p\right)^{1/p} \le C\left(\int_0^1 \left\|\sum_k r_k(t)x_k\right\|^2 dt\right)^{1/2};\qquad(2.71)$$

here r_k as usual stands for the ith Rademacher function on $[0,1]$. It is well-known that each $L_p(\mu)$ has cotype $\max\{p,2\}$. A Banach space X is said to have finite cotype if it has cotype p for some $2 \le p < \infty$.

Corollary 8. *Let $E(\mu)$ be a Banach function space and X a Banach space, both of finite cotype. Assume that $\sum_k x_k$ is an unconditionally convergent series in $E(\mu,X)$, and f its sum. Then there is a Riesz matrix $R^\lambda = (r^\lambda_{jk})$ such that*

$$\sup_j \left\| \sum_{k=0}^{\infty} r^\lambda_{jk} \sum_{\ell=0}^{k} x_\ell(\cdot)\right\|_X \in E(\mu),$$

and μ-almost everywhere

$$\lim_j \sum_{k=0}^{\infty} r^\lambda_{jk} \sum_{\ell=0}^{k} x_\ell = f.$$

In the case of orthonormal series this interesting result is a relatively simple consequence on Zygmund's work from [97] (see e.g. [1, p.142]).

Proof. It can be seen easily that $E(\mu,X)$ has finite cotype, say cotype r for $2 \le r < \infty$ (see e.g. [57, Theorem 3.3]). We know that the operator

$$u : c_0 \longrightarrow E(\mu,X), \; ue_k := x_k$$

by a result of Maurey is q-summing for each $r < q < \infty$; indeed, the fact that $E(\mu,X)$ has cotype r implies that u is $(r,1)$-summing, and then it is $r + \varepsilon$-summing for each $\varepsilon > 0$ (see e.g. [6, Sect. 24.7]). Fix such q. Then by (2.16) we get a factorization

where v is some operator and D_α is a diagonal operator with $\alpha \in \ell_q$. In particular, we see that $x_k = \alpha_k y_k$ where the $y_k := v(e_k)$ form a weakly q'-summable sequence in $E(\mu,X)$. Choose a positive sequence (μ_k) which increases to ∞ and which satisfies $\sum_k |\alpha_k \mu_k|^q < \infty$. Define first $\lambda_k := 2^{2^{\mu_k}}$, hence $\sum_k |\alpha_k \log\log \lambda_k|^q < \infty$, and second the desired Riesz matrix R^λ by

$$r^\lambda_{jk} := \begin{cases} \dfrac{\lambda_{k+1} - \lambda_k}{\lambda_{j+1}} & k \le j \\[2mm] 0 & k > j. \end{cases}$$

By Theorem 8 the matrix product $A = R^\lambda \, \Sigma \, D_{(1/\log\log\lambda_k)}$ given by

$$a_{jk} = \begin{cases} \left(1 - \dfrac{\lambda_k}{\lambda_{j+1}}\right) \dfrac{1}{\log\log\lambda_k} & k \le j \\[2mm] 0 & k > j \end{cases}$$

is (p,q)-maximizing – in particular, we have that

$$\sup_j \left| \sum_k r_{jk}^\lambda \sum_{\ell=0}^{k} x_\ell \right| = \sup_j \left| \sum_{k=0}^{\infty} a_{jk} \, \alpha_k \, \log\log\lambda_k \, y_k \right| \in E(\mu, X).$$

In order to obtain the second statement we conclude from Proposition 2 that

$$\left(\sum_k r_{jk}^\lambda \sum_{\ell=0}^{k} x_\ell \right)_j = \left(\sum_{k=0}^{\infty} a_{jk} \, \alpha_k \, \log\log\lambda_k \, y_k \right)_j$$

converges μ-almost everywhere. Since (r_{jk}^λ) is a summation process we finally see – taking the limit first in $E(\mu, X)$ – that

$$f = \sum_k x_k = \lim_j \sum_k r_{jk}^\lambda \sum_{\ell=0}^{k} x_\ell \qquad \mu - \text{a.e.},$$

which completes the proof.   □

2.3.2 *Laws of Large Numbers in Banach Function Spaces*

Given a sequence of random variables X_k on a probability space all with variation 0, a typical law of large numbers isolates necessary conditions under which the arithmetic means

$$\frac{1}{j+1} \sum_{k=0}^{j} X_k$$

converge to zero almost everywhere. Of course, theorems of this type also make sense if instead of the arithmetic means we take linear means

$$\sum_{k=0}^{j} s_{jk} \sum_{\ell=0}^{k} X_\ell$$

with respect to a given lower triangle summation process S. Via Kronecker's Lemma 11 each coefficient test for orthonormal series generates a law of large numbers for orthogonal sequences – this is the content of the following

Lemma 14. *Let S be an lower triangular summation method and ω a Weyl sequence. Then for each orthogonal sequence (x_k) in $L_2(\mu)$ with $\sum_k \frac{\omega_k^2}{k^2} \|x_k\|_2^2 < \infty$ we have*

$$\lim_j \frac{1}{j+1} \sum_{k=0}^{j} s_{jk} \sum_{\ell=0}^{k} x_\ell = 0 \quad \mu - a.e.$$

If S in addition satisfies that for each orthonormal series $\sum_k \alpha_k x_k$ in $L_2(\mu)$

$$\sup_j \left| \sum_{k=0}^{j} s_{jk} \sum_{\ell=0}^{k} \frac{\alpha_\ell}{\omega_\ell} x_\ell \right| \in L_2(\mu),$$

then for each orthogonal sequence (x_k) in $L_2(\mu)$ with $\sum_k \frac{\omega_k^2}{k^2} \|x_k\|_2^2 < \infty$

$$\sup_j \left| \frac{1}{j+1} \sum_{k=0}^{j} s_{jk} \sum_{\ell=0}^{k} x_\ell \right| \in L_2(\mu).$$

This result in particular applies to ordinary summation, Riesz$^\lambda$ summation, Cesàro^r summation or Abel$^\rho$ summation, and ω in this case is the related optimal Weyl sequence (see Sect. 2.2).

Proof. Take some orthogonal sequence (x_k) in $L_2(\mu)$ such that $\sum_k \frac{\omega_k^2}{k^2} \|x_k\|_2^2 < \infty$. Then $\sum_k \frac{\omega_k \|x_k\|_2}{k} \frac{x_k}{\|x_k\|_2}$ is an orthonormal sequence, and since ω is a Weyl sequence for S we see that

$$\lim_j \sum_{k=0}^{j} s_{jk} \sum_{\ell=0}^{k} \frac{x_\ell}{\ell} = \sum_{k=0}^{\infty} \frac{x_k}{k} \quad \mu - a.e.$$

Define the matrix $A = S\Sigma$ and note that for each choice of finitely many scalars $\xi_0, \dots, \xi_j$

$$\sum_{k=0}^{j} a_{jk} \xi_k = \sum_{k=0}^{j} s_{jk} \sum_{\ell=0}^{k} \xi_\ell.$$

Hence by Kronecker's Lemma 11,(2) we see that

$$0 = \lim_j \frac{1}{j+1} \sum_{k=0}^{j} a_{jk} x_k = \lim_j \frac{1}{j+1} \sum_{k=0}^{j} s_{jk} \sum_{\ell=0}^{k} x_\ell \quad \mu - a.e.$$

To prove the second result, note that by assumption we have that

$$\sup_j \left| \sum_{k=0}^{j} s_{jk} \sum_{\ell=0}^{k} \frac{x_\ell}{\ell} \right| \in L_2(\mu).$$

Hence we apply Lemma 11,(1) to conclude that

$$\sup_j \left| \frac{1}{j+1} \sum_{k=0}^{j} s_{jk} \sum_{\ell=0}^{k} x_\ell \right| = \sup_j \left| \sum_{k=0}^{j} \frac{1}{j+1} a_{jk} x_k \right|$$

$$\leq 2 \sup_j \left| \sum_{k=0}^{j} a_{jk} \frac{x_k}{k} \right|$$

$$= 2 \sup_j \left| \sum_{k=0}^{j} s_{jk} \sum_{\ell=0}^{k} \frac{x_\ell}{\ell} \right| \in L_2(\mu),$$

the conclusion.   □

To see an example we mention the law of large numbers which in the sense of the preceding result corresponds to the Menchoff-Rademacher theorem 6 (see e.g. [82, p.86-87]): *For each orthogonal system (x_k) in $L_2(\mu)$ with $\sum_k \frac{\log^2 k}{k^2} \|x_k\|_2^2 < \infty$ we have*

$$\lim_j \frac{1}{j+1} \sum_{k=0}^{j} x_k = 0 \quad \mu - a.e.$$

and

$$\sup_j \left| \frac{1}{j+1} \sum_{k=0}^{j} x_k \right| \in L_2(\mu). \tag{2.72}$$

The main aim of this section is to show that each such law of large numbers for orthogonal sequences of square integrable random variables which additionally satisfies a maximal inequality like in (2.72), transfers in a very complete sense to a law of large numbers in vector-valued Banach function spaces $E(\mu, X)$.

Theorem 15. *Let S be a lower triangular summation method. Assume that ω is an increasing sequence of positive scalars such that for each orthogonal sequence (x_k) in $L_2(\mu)$ with $\sum_k \frac{\omega_k^2}{k^2} \|x_k\|_2^2 < \infty$ we have*

$$\sup_j \left| \frac{1}{j+1} \sum_{k=0}^{j} s_{jk} \sum_{\ell=0}^{k} x_\ell \right| \in L_2(\mu).$$

Then for each unconditionally convergent series $\sum_k \frac{\omega_k}{k} x_k$ in $E(\mu, X)$

(1) $\sup_j \left\| \frac{1}{j+1} \sum_{k=0}^{j} s_{jk} \sum_{\ell=0}^{k} x_\ell(\cdot) \right\|_X \in E(\mu)$

(2) $\lim_j \frac{1}{j+1} \sum_{k=0}^{j} s_{jk} \sum_{\ell=0}^{k} x_\ell = 0 \quad \mu\text{-a.e.}$

Proof. We have to repeat part of the preceding proof. For every orthonormal series $\sum_k \alpha_k x_k$ in $L_2(\mu)$ we have by assumption that

$$\sup_j \left| \sum_{k=0}^{j} \frac{1}{j+1} s_{jk} \sum_{\ell=0}^{k} \frac{\ell}{\omega_\ell} \alpha_\ell x_\ell \right| \in L_2(\mu).$$

Moreover for

$$
b_{jk} := \begin{cases} \dfrac{k}{(j+1)\omega_k}\sum_{\ell=k}^{j} s_{j\ell} & k \le j \\ 0 & k > j \end{cases}
$$

(compare with (2.5)) we have for each choice of scalars $\xi_0, \ldots, \xi_j$ that

$$
\sum_{k=0}^{j} b_{jk}\xi_k = \sum_{k=0}^{j} \frac{1}{j+1} s_{jk} \sum_{\ell=0}^{k} \frac{\ell}{\omega_\ell} \xi_\ell .
$$

Hence, we deduce from Theorem 1 that B is $(2,2)$-maximizing, and obtain (1) from Theorem 13, (1). Moreover, since the kth column of B converges to 0 (compare with (2.67)), we deduce from Theorem 13, (2) that the limit in (2) exists almost everywhere, and it remains to show that this limit is 0 almost everywhere. Define the matrix $A = S\Sigma$. Since S is a summation method and the series $\sum_k \frac{x_k}{k}$ converges in $E(\mu, X)$, we have

$$
\sum_{k=0}^{\infty} \frac{x_k}{k} = \lim_j \sum_{k=0}^{j} s_{jk} \sum_{\ell=0}^{k} \frac{x_\ell}{\ell} = \lim_j \sum_{k=0}^{j} a_{jk} \frac{x_k}{k},
$$

the limits taken in $E(\mu, X)$. Hence by Kronecker's Lemma 11,(2) we see that in $E(\mu, X)$

$$
0 = \lim_j \frac{1}{j+1}\sum_{k=0}^{j} a_{jk} x_k = \lim_j \frac{1}{j+1}\sum_{k=0}^{j} s_{jk} \sum_{\ell=0}^{k} x_\ell .
$$

As a consequence a subsequence of the latter sequence converges almost everywhere to 0 which clearly gives the claim.   □

As a particular case, we deduce from (2.72) the following

Corollary 9. *For sequences (x_k) in $E(\mu, X)$ for which $\sum_k \frac{\log k}{k} x_k$ converges unconditionally we have*

$$
\lim_j \frac{1}{j+1}\sum_{k=0}^{j} x_k = 0 \quad \mu - a.e.
$$

and

$$
\sup_j \left\| \frac{1}{j+1}\sum_{k=0}^{j} x_k(\cdot) \right\|_X \in E(\mu).
$$

Applying Theorem 9 to Theorem 15 we obtain in the same way that, given a sequence (x_k) in $E(\mu, X)$ for which $\sum_k \frac{\log\log k}{k} x_k$ converges unconditionally, we have

$$
\lim_j \frac{1}{(j+1)^2}\sum_{k=0}^{j}\sum_{\ell=0}^{k} x_\ell = 0 \quad \mu - a.e. \tag{2.73}
$$

and

$$\sup_j \left\| \frac{1}{(j+1)^2} \sum_{k=0}^{j} \sum_{\ell=0}^{k} x_\ell(\cdot) \right\|_X \in E(\mu).$$

It is surprising and part of the next theorem that in contrast to the situation in Corollary 9 the double logarithmic term in the assumption for (2.73) is superfluous – even for Cesàro summation of arbitrary $r > 0$.

Theorem 16. *Let $\sum_k \frac{x_k}{k}$ be an unconditionally convergent series in some vector-valued Banach function space $E(\mu,X)$. Then for each $r > 0$ we have*

(1) $\sup_j \left\| \frac{1}{j+1} \sum_{k=0}^{j} \frac{A_{j-k}^{r-1}}{A_j^r} \sum_{\ell=0}^{k} x_\ell \right\|_X \in E(\mu)$

(2) $\lim_j \frac{1}{j+1} \sum_{k=0}^{j} \frac{A_{j-k}^{r-1}}{A_j^r} \sum_{\ell=0}^{k} x_\ell = 0 \quad \mu - a.e.$

For the very special case of orthogonal sequences (x_k) in some $L_2(\mu)$ statement (2) of this result is due to Moricz [63, Theorem 2]; our proof will use Theorem 11 which after all was a consequences of the maximal inequalities (2.54) and (2.59).

Proof. Recall the definition of Cesàro summation of order r from Sect. 2.1.1:

$$c_{jk}^r := \begin{cases} \dfrac{A_{j-k}^{r-1}}{A_j^r} & k \le j \\ 0 & k > j, \end{cases}$$

and that for each choice of scalars $\xi_0, \dots, \xi_j$ we have

$$\frac{1}{j+1} \sum_{k=0}^{j} \frac{A_{j-k}^{r-1}}{A_j^r} \sum_{\ell=0}^{k} \ell \xi_\ell = \sum_{k=0}^{j} \frac{k}{j+1} \frac{A_{j-k}^r}{A_j^r} \xi_k$$

(see (2.45)). Moreover, we proved in Theorem 11 that the matrix M^r defined by

$$m_{jk}^r := \begin{cases} \dfrac{k}{j+1} \dfrac{A_{j-k}^r}{A_j^r} & k \le j \\ 0 & k > j \end{cases}$$

is $(2,2)$-maximizing. Hence, we know that by the very definition of maximizing matrices for each orthogonal sequence (x_k) in $L_2(\mu)$ with $\sum_k \frac{\|x_k\|_2^2}{k^2} < \infty$ we have

$$\sup_j \left| \frac{1}{j+1} \sum_{k=0}^{j} \frac{A_{j-k}^{r-1}}{A_j^r} \sum_{\ell=0}^{k} x_\ell \right| \in L_2(\mu),$$

i.e. the matrix C^r satisfies the assumptions of Theorem 15 which in turn gives the desired result ($\omega_k = 1$). □

Chapter 3
Noncommutative Theory

3.1 The Tracial Case

We repeat and extend our program from the second chapter within the setting of symmetric spaces $E(\mathcal{M}, \tau)$ of operators constructed over semifinite integration spaces $(\mathcal{M}, \tau)$ (a von Neumann algebra $\mathcal{M}$ together with a normal, faithful and semifinite trace τ) and symmetric Banach function spaces E (realized in $L_0[0, \tau(1))$). Most of this material developed from the ideas given in [7].

The two main results of Sect. 2.3 on limit theorems in vector-valued Banach function spaces were Theorems 13 and 14. According to the first theorem, for any $(2,2)$-maximizing matrix (a_{jk}) and each unconditionally summable sequence (x_k) in some $E(\mu, X)$ the maximal function $\sup_j \| \sum_j a_{jk} x_k(\cdot) \|_X$ belongs to $E(\mu)$, and the second one then extends this result to arbitrary (p,q)-maximizing matrices provided $E(\mu)$ satisfies a certain geometric condition and the (x_k) are taken from a more restrictive class of sequences in $E(\mu, X)$. In the Corollaries 6 and 7 both theorems are applied to ordinary summation as well as Cesàro, Riesz and Abel summation – this leads to coefficient tests with and without log-terms.

The main aim of this section is to show that in these theorems and their applications the vector-valued Banach function spaces $E(\mu, X)$ can be replaced by symmetric spaces $E(\mathcal{M}, \tau)$ of operators. It turns out that our theory of maximizing matrices is suitable to convert major parts of the commutative theory into a theory on summation of unconditionally convergent series in noncommutative symmetric spaces $E(\mathcal{M}, \tau)$ of operators. Our substitutes of the above mentioned commutative results now are the noncommutative extension theorems 17, 18, and 19 for maximizing matrices. Applied to abstract summation methods these results yield the Theorems 20 and 21 which applied to ordinary, Cesàro, Riesz and Abel sumation give the Corollaries 10 and 11.

A. Defant, *Classical Summation in Commutative and Noncommutative Lp-Spaces*, Lecture Notes in Mathematics 2021, DOI 10.1007/978-3-642-20438-8_3,
© Springer-Verlag Berlin Heidelberg 2011

3.1.1 Symmetric Spaces of Operators

Here we collect some relevant facts on noncommutative L_p-spaces $L_p(\mathcal{M}, \tau)$ constructed over a noncommutative integration space $(\mathcal{M}, \tau)$, where $\mathcal{M}$ is a semifinite von Neumann algebra acting on a Hilbert space H and τ a faithful, normal and semifinite trace. These Banach spaces are all contained in the metrizable space $L_0(\mathcal{M}, \tau)$ of all τ-measurable (unbounded) operators affiliated with $\mathcal{M}$ which provides a very comfortable setting for our study of almost everywhere summation of unconditionally convergent series in such spaces. More generally, we consider symmetric spaces of operators in $L_0(\mathcal{M}, \tau)$ constructed over a noncommutative integration space $(\mathcal{M}, \tau)$ and a symmetric Banach function space E of functions defined on the real interval $[0, \tau(1))$.

Von Neumann algebras. Recall that a von Neumann algebra $\mathcal{M}$ of operators acting on a Hilbert space H is a unitial C^*-subalgebra of $\mathcal{L}(H)$ which is closed in the strong operator topology or equivalently weak operator topology; we write $(\cdot|\cdot)$ for the scalar product on H. It is well-known that a C^*-subalgebra $\mathcal{M} \subset \mathcal{L}(H)$ is a von Neumann algebra if and only if its bicommutant $\mathcal{M}^{\sharp\sharp}$ equals $\mathcal{M}$ if and only if it has a (unique) predual $\mathcal{M}_*$ as a Banach space. The identity in $\mathcal{M}$ is denoted by 1, and $\mathcal{M}_{\text{proj}}$ stands for all (orthogonal) projections in $\mathcal{M}$. Recall that for two such projections we have $p \leq q$ whenever $pH \subset qH$, and that for every family $(p_i)_I$ of projections $\inf_I p_i$ is the projection onto $\cap_I p_i H$ and $\sup_I p_i$ the projection onto $\overline{\cup_I p_i H}$. A functional $\tau : \mathcal{M}_{\geq 0} \to [0, \infty]$ is said to be a trace if it is additive, homogeneous with respect to positive scalars and satisfies the condition

$$\tau(xx^*) = \tau(x^*x) \text{ for all } x \in \mathcal{M}_{\geq 0};$$

moreover, it is normal whenever $x_\alpha \uparrow x$ in $\mathcal{M}_{\geq 0}$ (strong operator topology) assures that $\tau(x_\alpha) \to \tau(x)$, it is faithful if $\tau(x) = 0$ implies $x = 0$, and semifinite if for every $x \in \mathcal{M}_{\geq 0}$ there is $0 \neq y \in \mathcal{M}_{\geq 0}$ such that $y \leq x$ and $\tau(y) < \infty$.

Throughout the rest of this section we fix a noncommutative integration space

$$(\mathcal{M}, \tau),$$

i.e. a (semifinite) von Neumann algebra $\mathcal{M}$ of operators acting on a Hilbert space H and a normal, faithful and semifinite trace $\tau : \mathcal{M}_{\geq 0} \to [0, \infty]$. We often think of τ as a noncommutative quantum probability. It is well-known that each σ-finite measure space (Ω, Σ, μ) defines such a pair: $\mathcal{M} = L_\infty(\mu)$ and $\tau(f) = \int f d\mu$.
References: See e.g. [37, 44, 65, 89].

Unbounded operators. We have to consider unbounded operators a on H with domain $\text{dom}(a)$ and range $\text{range}(a)$. By $r(a)$ we denote the projection onto $\overline{\text{range}}(a)$, the range projection of a. Clearly, $r(a) = \inf p$, the infimum taken over all projections p with $pa = a$.

Lemma 15. *Let a be a selfadjoint operator on H and p a projection on H with $pa = a$. Then $a = pa = ap$, and the restriction of a to pH is selfadjoint. Moreover, if $p = r(a)$, then this restriction is injective.*

For the sake of completeness we give a proof.

Proof. Obviously, $ap = a^*p = (pa)^* = a$ (note that p is bounded), and hence the restriction of a to pH has dense domain and is again selfadjoint. Let us prove that for $p = r(a)$ this restriction is injective: Assume that there is $0 \neq \xi_0 \in \mathrm{dom}(a) \cap pH$ with $a(\xi_0) = 0$. Let p_0 be the projection on H onto the (orthogonal) complement of $\mathrm{span}\xi_0$ in pH, hence $pH = \mathrm{span}\xi_0 \oplus p_0 H$. Obviously, $p_0 \leq p$ and $p_0 \neq p$. We prove that $a = ap_0$ which then implies that $p_0 a = a$, a contradiction. We have to show that $\mathrm{dom}(a) = \mathrm{dom}(ap_0)$ and that $a\xi = ap_0\xi$ for each $\xi \in \mathrm{dom}(a)$. Take $\xi \in \mathrm{dom}(a) = \mathrm{dom}(ap)$. Then $p\xi \in \mathrm{dom}(a)$ which gives for some λ that $p_0\xi = p\xi - \lambda\xi_0 \in \mathrm{dom}(a)$, and therefore $ap_0\xi = a(p\xi - \lambda\xi_0) = ap\xi = a\xi$. Conversely, for $\xi \in \mathrm{dom}(ap_0)$ (i.e. $p_0\xi \in \mathrm{dom}(a)$) we have that $p\xi = \lambda\xi_0 + p_0\xi \in \mathrm{dom}(a)$, hence $\xi \in \mathrm{dom}(ap) = \mathrm{dom}(a)$. This completes the argument. □

Recall that two positive and selfadjoint operators a and b fulfill the relation $a \leq b$ if $\mathrm{dom}(b^{1/2}) \subset \mathrm{dom}(a^{1/2})$ and $(a^{1/2}\xi | a^{1/2}\xi) \leq (b^{1/2}\xi | b^{1/2}\xi)$ for all $\xi \in \mathrm{dom}(b^{1/2})$. The following remark will be needed.

Lemma 16. *Let x and y be two positive and selfadjoint operators on a Hilbert space H with $x \leq y$, y injective. Then $y^{1/2}$ is injective and $x^{1/2}y^{-1/2}$ is a contraction on H.*

Proof. Clearly, $y^{1/2}$ is injective. Since it is also selfadjoint, its inverse $y^{-1/2}$ is a selfadjoint operator with domain $\mathrm{range}(y^{1/2})$ and range $\mathrm{dom}(y^{1/2})$. Observe that $u := x^{1/2}y^{-1/2}$ has dense domain: Since $\mathrm{dom}(y^{1/2}) \subset \mathrm{dom}(x^{1/2})$, we have that $\mathrm{dom}(u) = \{\xi \in \mathrm{dom}(y^{-1/2}) \, | \, y^{-1/2}\xi \in \mathrm{dom}(x^{1/2})\} = \mathrm{range}(y^{1/2})$. It now suffices to check that u is a contraction on $\mathrm{dom}(u)$. But this follows easily from the fact that $y^{-1/2}\xi \in \mathrm{dom}(y^{1/2})$ for each $\xi \in \mathrm{dom}(u)$, and hence

$$(u\xi | u\xi) = (x^{1/2}y^{-1/2}\xi | x^{1/2}y^{-1/2}\xi) \leq (y^{1/2}y^{-1/2}\xi | y^{1/2}y^{-1/2}\xi) = (\xi | \xi),$$

the conclusion. □

If x is any selfadjoint (densely defined and closed) operator on H, then

$$x = \int_{\mathbb{R}} \lambda \, de_\lambda^x,$$

where e_λ^x denotes the spectral measure defined by x. We write $\chi_B(x)$ for the spectral projection $\int_B de_\lambda^x$, $B \subset \mathbb{R}$ a Borel set. In particular, we note that $e_\lambda^x = \chi_{(-\infty,\lambda]}(x)$. References: See e.g. [44, 90, 93].

Measurable operators. A closed and densely defined operator a on H is said to be affiliated with $\mathcal{M}$ if $ya \subset ay$ (i.e. ay extends ya) for all y in the commutant $\mathcal{M}^\sharp$

of $\mathcal{M}$. An operator a affiliated with $\mathcal{M}$ is said to be τ-measurable if for every $\delta > 0$ there is a projection $p \in \mathcal{M}_{\text{proj}}$ such that $\tau(1-p) \leq \delta$ and $ap \in \mathcal{M}$. Let

$$L_0(\mathcal{M}, \tau)$$

be the set of all τ-measurable operators on H which together with the respective closures of the algebraic sum and product forms a $*$-algebra. The sets $N(\varepsilon, \delta)$, consisting of all $a \in L_0(\mathcal{M}, \tau)$ for which there is $p \in \mathcal{M}_{\text{proj}}$ such that $ap \in \mathcal{M}$ with $\|ap\| \leq \varepsilon$ and $\tau(1-p) \leq \delta$, form a 0-neighbourhood basis for a metrizable linear topology on $L_0(\mathcal{M}, \tau)$, called the measure topology. This way $L_0(\mathcal{M}, \tau)$ becomes a complete metrizable topological $*$-algebra.

A general philosophy is that the operators in $L_0(\mathcal{M}, \tau)$ "behave almost like" those in $\mathcal{M}$. For example, $a = a^{**}$ and $(ab)^* = b^* a^*$ for all $a, b \in L_0(\mathcal{M}, \tau)$, $a \in L_0(\mathcal{M}, \tau)$ is selfadjoint whenever it is positive, the range projection of $a \in L_0(\mathcal{M}, \tau)$ belongs to $\mathcal{M}_{\text{proj}}$, and any $a \in L_0(\mathcal{M}, \tau)$ has a polar decomposition $a = u|a|$ where $|a| = (a^* a)^{1/2}$ and u is a partial isometry in $\mathcal{M}$. We also recall that the trace τ on $\mathcal{M}$ can be extended to all positive operators $x \in L_0(\mathcal{M}, \tau)$ through the definition

$$\tau(x) := \sup_n \tau\left(\int_0^n \lambda\, de_\lambda^x\right) \in [0, \infty].$$

Finally, note that $x \leq y$ for $0 \leq x, y \in L_0(\mathcal{M}, \tau)$ (see the above definition) if and only if $(x\xi|\xi) \leq (y\xi|\xi)$ for all $\xi \in \text{dom}(x) \cap \text{dom}(y)$.

To see examples, note that for $\mathcal{M} = \mathscr{L}(H)$ together with the standard trace tr, $L_0(\mathcal{M}, \tau)$ coincides with $\mathcal{M}$, and the measure topology is the norm topology. Or if $\tau(1) < \infty$, then $L_0(\mathcal{M}, \tau)$ equals the space of all operators affiliated with $\mathcal{M}$. Moreover, in the commutative case $\mathcal{M} = L_\infty(\mu)$ and $\tau(f) = \int f\, d\mu$ the space $L_0(\mathcal{M}, \tau)$ is nothing else than the closure of $L_\infty(\mu)$ in $L_0(\mu)$, all measurable functions on Ω endowed with the usual topology of convergence in measure. References: See e.g. [37, 44, 66, 90].

Decreasing rearrangement. For each operator x on H affiliated with $\mathcal{M}$, all spectral projections $\chi_B(|x|)$ belong to $\mathcal{M}$, and $x \in L_0(\mathcal{M}, \tau)$ if and only if $\tau(\chi_{(\lambda, \infty)}(|x|)) < \infty$ for some $\lambda \in \mathbb{R}$. Recall the definition of the decreasing rearrangement of an operator $x \in L_0(\mathcal{M}, \tau)$: For $t > 0$

$$\mu_t(x) := \inf\left\{\lambda > 0 \,\big|\, \tau(\chi_{(\lambda, \infty)}(|x|)) \leq t\right\},$$

and an equivalent description is given by

$$\mu_t(x) = \inf\left\{\|xp\|_\infty \,\big|\, p \in \mathcal{M}_{\text{proj}} \text{ with } \tau(1-p) \leq t\right\}. \tag{3.1}$$

The function $\mu_t(x)$ in the variable t is briefly denoted by $\mu(x)$; clearly, in the commutative situation $\mathcal{M} = L_\infty(\nu)$ and $\tau(f) = \int f\, d\nu$ the decreasing rearrangement of any $f \in L_0(\mathcal{M}, \tau)$ (the closure of $L_\infty(\nu)$ in $L_0(\nu)$) is nothing else than the classical decreasing rearrangement of the function $|f|$. An important fact is that as in the commutative case we have

$$\tau(x) = \int_{\mathbb{R}} \mu_t(x)\,dt\,,\ x \in L_0(\mathcal{M},\tau).$$

If $(\mathcal{M},\tau) = (\mathcal{L}(H),\mathrm{tr})$, all bounded linear operators on H equipped with the standard trace tr, then for any compact $x \in L_0(\mathcal{M},\tau) = \mathcal{L}(H)$ the decreasing rearrangement $\mu(x)$ in a natural way may be identified with the sequence of singular values of $|x| = \sqrt{x^*x}$, repeated according to multiplicity and arranged in decreasing order.

References: See e.g. [16, 17, 22].

Symmetric spaces of operators. We now recall the definition of a symmetric operator space $E(\mathcal{M},\tau)$ buildup with respect to a noncommutative measure space $(\mathcal{M},\tau)$ and a symmetric Banach function space. Fix some $\alpha \in \mathbb{R}_{>0} \cup \{\infty\}$. A Banach function space $(E,\|\cdot\|_E)$ of functions in $L_0[0,\alpha)$ (see Sect. 2.1.2 for the definition) is said to be rearrangement invariant (on the interval $[0,\alpha)$) whenever for $f \in L_0[0,\alpha)$ and $g \in E$ the inequality $\mu(f) \leq \mu(g)$ assures that $f \in E$ and $\|f\|_E \leq \|g\|_E$. Such a rearrangement invariant Banach function space E is said to be symmetric provided $f \prec\prec g$ for $f,g \in E$ implies that $\|f\|_E \leq \|g\|_E$; here $f \prec\prec g$ as usual denotes the submajorization in the sense of Hardy-Littlewood-Polya: for all $t > 0$

$$\int_0^t \mu_s(f)\,ds \leq \int_0^t \mu_s(g)\,ds.$$

To see examples, L_p-, Orlicz, Lorentz and Marcinkiewicz spaces are rearrangement invariant Banach function spaces. Moreover, each rearrangement invariant Banach function space E with a Fatou norm $\|\cdot\|_E$ (i.e. $0 \leq f_\alpha \uparrow f \in E$ implies $\|f_\alpha\|_E \uparrow \|f\|_E$) is symmetric.

Given a semifinite von Neumann algebra $\mathcal{M}$ together with a semifinite, faithful and normal trace τ, and given a symmetric Banach function space $(E,\|\cdot\|_E)$ on the interval $[0,\tau(1))$ the vector space

$$E(\mathcal{M},\tau) := \left\{ x \in L_0(\mathcal{M},\tau)\,\big|\,\mu(x) \in E \right\}$$

together with the norm

$$\|x\|_{E(\mathcal{M},\tau)} := \|\mu(x)\|_E$$

forms a Banach space, here called the symmetric space of operators with respect to $(\mathcal{M},\tau)$ and E. An important fact is that the following two natural contractions hold true:

$$L_1(\mathcal{M},\tau) \cap L_\infty(\mathcal{M},\tau) \hookrightarrow E(\mathcal{M},\tau) \hookrightarrow L_1(\mathcal{M},\tau) + L_\infty(\mathcal{M},\tau); \qquad (3.2)$$

here as usual the intersection is endowed with the maximum norm and the sum with the sum norm.

We remark that Kalton and Sukochev in [45] proved that this construction even then leads to a Banach space $E(\mathcal{M},\tau)$ if the function space E is only assumed to be rearrangement invariant (in other words, no submajorization is needed in order to prove the triangle inequality and completeness of $\|\mu(\cdot)\|_E$).

References: See e.g. [11, 12, 13, 14, 45, 53, 54, 69, 70, 80, 85, 86, 87, 88, 96].

Noncommutative L_p-spaces. For $1 \leq p < \infty$ and $E = L_p[0,\tau(1))$ we have that $E(\mathcal{M},\tau)$ coincides with the noncommutative L_p-spaces

$$L_p(\mathcal{M},\tau) = \left\{ x \in L_0(\mathcal{M},\tau) \,\big|\, \|x\|_p := \tau(|x|^p)^{1/p} < \infty \right\};$$

as usual, we put

$$L_\infty(\mathcal{M},\tau) := \mathcal{M},$$

and write from now on $\|\cdot\|_\infty$ for the norm in $\mathcal{M}$. It is well-known that non-commutative L_p-spaces satisfy all the expected properties. We for example have that $x \in L_p(\mathcal{M},\tau)$ if and only if $x^* \in L_p(\mathcal{M},\tau)$, and $\|x\|_p = \|x^*\|_p$. Or for $1/r = 1/p + 1/q$ the usual Hölder's inequality extends to the noncommutative setting: For $x \in L_p(\mathcal{M},\tau)$ and $y \in L_q(\mathcal{M},\tau)$

$$\|xy\|_r \leq \|x\|_p \|y\|_q.$$

This defines a natural duality between $L_p(\mathcal{M},\tau)$ and $L_q(\mathcal{M},\tau)$ through the duality bracket $<x,y> = \tau(xy)$,

$$L_p(\mathcal{M},\tau)' = L_q(\mathcal{M},\tau) \text{ isometrically};$$

in particular, $L_1(\mathcal{M},\tau)$ is the predual $\mathcal{M}_*$ of $\mathcal{M}$.

Clearly, the construction of symmetric spaces of operators applied to the commutative integration space build by some $L_\infty(\mu)$ and its integral as trace, leads to the classical spaces $L_p(\mu)$. The integration spaces given by $\mathscr{L}(H)$, all (bounded and linear) operators on some Hilbert space H in combination with the canonical trace $\tau = \mathrm{tr}$ on this space, generates the Schatten classes S_p, and in a similar way unitary ideals S_E can be viewed as symmetric spaces of operators generated by $(\mathscr{L}(H),\mathrm{tr})$ and a symmetric Banach sequence space E. For the definition of the hyperfinite factor together with its canonical trace see Sect. 3.1.8. Of course, many more examples can be found in the following references (and the ones given on symmetric spaces of operators).

References: See e.g. [37, 66, 80, 90].

Powers of symmetric spaces of operators. In order to define powers of symmetric spaces of operators recall that for $0 < r \leq \infty$ a Banach function space $E(\mu)$ (with respect to some measure space (Ω,μ)) is said to be r-convex (already explained in (2.68)) and r-concave, respectively, whenever there is a constant $C \geq 0$ such that for each choice of finitely many $x_1,\ldots,x_n \in E$

$$\left\|\left(\sum_{k=1}^{n}|x_k|^r\right)^{1/r}\right\|_E \leq C\left(\sum_{k=1}^{n}\|x_k\|_E^r\right)^{1/r},$$

and

$$\left(\sum_{k=1}^{n}\|x_k\|_E^r\right)^{1/r} \leq C\left\|\left(\sum_{k=1}^{n}|x_k|^r\right)^{1/r}\right\|_E,$$

respectively; as usual, the best constant $C \geq 0$ is denoted by $M^{(r)}(E)$ resp. $M_{(r)}(E)$. We recall that for $r_1 \leq r_2$

$$M^{(r_1)}(E) \leq M^{(r_2)}(E) \text{ and } M_{(r_2)}(E) \leq M_{(r_1)}(E).$$

To see an example: Each $L_p(\mu)$ is p-convex and p-concave with constants 1, and as a consequence $M^{(2)}(L_p(\mu)) = 1$ for $2 \leq p$ and $M_{(2)}(L_p(\mu)) = 1$ for $p \leq 2$. We will also use the fact that every r-convex (resp. r-concave) Banach function space may be renormed in such a way that its r-convexity (resp. r-concavity) constant is 1. For all needed information on convexity and concavity we once again refer to [53, 54].

If $M^{\max(1,r)}(E) = 1$, then the rth power

$$E^r(\mu) := \left\{x \in L_0(\mu) \mid |x|^{1/r} \in E(\mu)\right\}$$

endowed with the norm

$$\|x\|_{E^r} := \||x|^{1/r}\|_E^r$$

is again a Banach function space which is $1/\min(1,r)$-convex. Since for each operator $x \in L_0(\mathcal{M},\tau)$

$$\mu(|x|^r) = \mu(x)^r,$$

we conclude for every symmetric Banach function space E on the interval $[0, \tau(1))$ which satisfies $M^{\max(1,r)}(E) = 1$ that

$$E^r(\mathcal{M},\tau) = \left\{x \in L_0(\mathcal{M},\tau) \mid |x|^{1/r} \in E(\mathcal{M})\right\}$$

and

$$\|x\|_{E^r(\mathcal{M})} = \|\mu(|x|)\|_{E^r} = \|\mu(|x|^{1/r})\|_E^r = \||x|^{1/r}\|_{E(\mathcal{M})}^r. \tag{3.3}$$

References: See e.g. [13, 14, 16, 17, 37, 96].

Köthe duality of symmetric spaces of operators. We also need to recall the notion of Köthe duals of symmetric spaces of operators. As in the commutative case (given a symmetric Banach function space E)

$$E(\mathcal{M},\tau)^{\times} := \left\{x \in L_0(\mathcal{M},\tau) \mid xy \in L_1(\mathcal{M},\tau) \text{ for all } y \in E(\mathcal{M},\tau)\right\}$$

together with the norm

$$\|x\|_{E(\mathcal{M})^\times} := \sup\left\{\tau(|xy|) \mid \|y\|_{E(\mathcal{M})} \leq 1\right\}$$

defines a Banach space – the Köthe dual of $E(\mathcal{M},\tau)$. It is known that for any ordercontinuous Banach function space E the following Banach space equality holds:

$$E(\mathcal{M},\tau)^\times = E^\times(\mathcal{M},\tau) = E(\mathcal{M},\tau)'. \tag{3.4}$$

Finally, we note that for $x,y \in L_0(\mathcal{M})$ with $xy, yx \in L_1(\mathcal{M})$ we have

$$\tau(xy) = \tau(yx). \tag{3.5}$$

In particular, this equality holds for $x \in E(\mathcal{M})$ and $y \in E^\times(\mathcal{M})$. If in this case both operators are even positive, then

$$\tau(xy) = \tau(x^{1/2}yx^{1/2}) = \tau(y^{1/2}xy^{1/2}). \tag{3.6}$$

References: See e.g. [14, 16, 17, 96].

3.1.2　Maximal Inequalities in Symmetric Spaces of Operators

A straightforward translation of the notion of a maximal function $\sup_n x_n$ for a sequence of functions (x_n) from a Banach function space $E(\mu)$ over a (σ-finite) measure spaces (Ω, Σ, μ) to sequences (x_n) of operators in a symmetric space $E(\mathcal{M},\tau)$ over a noncommutative integration space $(\mathcal{M},\tau)$ is not possible. Even for sequences of positive operators x_n there may be no operator x such that $(\xi|x\xi) = \sup_n(\xi|x_n\xi)$ for all $\xi \in H$.

Example 1. Consider the 2×2 matrices

$$a_1 = \begin{pmatrix} 2 & 0 \\ 0 & 0 \end{pmatrix}, \quad a_2 = \begin{pmatrix} 1 & 1 \\ 1 & 1 \end{pmatrix}, \quad a_3 = \begin{pmatrix} 0 & 0 \\ 0 & 1 \end{pmatrix}$$

Then there is no 2×2 matrix a such that for all $\xi \in \ell_2^2$

$$(\xi|a\xi) = \sup\left\{(\xi|a_1\xi), (\xi|a_2\xi), (\xi|a_3\xi)\right\}.$$

However as pointed out in Lemma 1, in the commutative situation a maximal function $\sup_n |x_n|$ belongs to $E(\mu)$ if and only if $(x_n) \in E(\mu)[\ell_\infty]$, i.e. there is a uniform factorization $x_n = y_n b$, where $b \in E(\mu)$ and (y_n) is a bounded sequence in $L_\infty(\mu)$. As observed by Pisier in [79] (for the hyperfinite case) and later by Junge in

[39] (for the general case) this algebraic formulation admits several noncommutative extensions. In complete analogy to Sect. 2.1.2 we now define ℓ_∞- and c_0-valued symmetric spaces of operators – objects which are fundamental for this chapter:

- $E(\mathcal{M})[\ell_\infty]$, the symmetric version
- $E(\mathcal{M})[\ell_\infty^c]$, the column version
- $E(\mathcal{M})[\ell_\infty^r]$, the row version
- $E(\mathcal{M})[\ell_\infty^{r+c}]$, the row+column version.

For their definitions fix again a noncommutative integration space $(\mathcal{M}, \tau)$, a von Neumann algebra together with a normal, faithful, semifinite trace τ, and moreover a symmetric rearrangement invariant Banach function space E on the interval $[0, \tau(1))$ and a countable partially ordered index set I. Define

$$E[\ell_\infty] = E(\mathcal{M})[\ell_\infty] = E(\mathcal{M}, \tau)[\ell_\infty(I)]$$

to be the space of all families $(x_n)_{n \in I}$ in $E(\mathcal{M}, \tau)$ which admit a factorization of the following form: There are $a, b \in E^{1/2}(\mathcal{M}, \tau)$ and a bounded family (y_n) in $\mathcal{M}$ such that for all n we have

$$x_n = a y_n b;$$

put

$$\|(x_n)\|_{E[\ell_\infty]} := \inf \|a\|_{E^{1/2}(\mathcal{M})} \sup_n \|y_n\|_\infty \|b\|_{E^{1/2}(\mathcal{M})},$$

the infimum taken over all possible factorizations. Moreover, let us define

$$E[\ell_\infty^c] = E(\mathcal{M})[\ell_\infty^c] = E(\mathcal{M}, \tau)[\ell_\infty^c(I)]$$

(here c should remind on the word "column") as the space of all $(x_n)_{n \in I}$ in $E(\mathcal{M}, \tau)$ for which there are $b \in E(\mathcal{M}, \tau)$ and a bounded sequence (y_n) in $\mathcal{M}$ such that $x_n = y_n b$, and

$$\|(x_n)\|_{E[\ell_\infty^c]} := \inf \sup_n \|y_n\|_\infty \|b\|_{E(\mathcal{M})}.$$

Similarly, we define the row version

$$E[\ell_\infty^r] = E(\mathcal{M})[\ell_\infty^r] = E(\mathcal{M}, \tau)[\ell_\infty^r(I)],$$

all sequences which allow a uniform factorization $x_n = a y_n$, again with $a \in E(\mathcal{M}, \tau)$ and (y_n) uniformly bounded in $\mathcal{M}$. Finally, let

$$E[\ell_\infty^{r+c}] = E(\mathcal{M})[\ell_\infty^{r+c}] = E(\mathcal{M}, \tau)[\ell_\infty^{r+c}(I)]$$

be the space of all families $x = (x_n)_{n \in I}$ in $E(\mathcal{M}, \tau)$ which can be written as a sum $x = u + v$ of two sequences $u \in E(\mathcal{M})[\ell_\infty^r]$ and $v \in E(\mathcal{M})[\ell_\infty^c]$; put

$$\|x\|_{E[\ell_\infty^{r+c}]} := \inf \|u\|_{E[\ell_\infty^r]} + \|v\|_{E[\ell_\infty^c]},$$

the infimum taken over all possible decompositions $x = u + v$.

The following standard lemma is needed to show that $E(\mathcal{M})[\ell_\infty]$, as well as its column and row version under an additional assumption on E, form (quasi) Banach spaces – we prove this lemma for the sake of completeness.

Lemma 17. *Let $c, c_1 \in L_0(\mathcal{M}, \tau)$ be positive with $c_1^2 \leq c^2$. Then there is a contraction $a_1 \in \mathcal{M}$ which satisfies $c_1 = a_1 c$. Moreover, if $c_1, c_2 \in L_0(\mathcal{M}, \tau)$ and $c := (c_1^2 + c_2^2)^{1/2}$, then there are contractions $a_1, a_2 \in \mathcal{M}$ such that*

$$c_k = a_k c \text{ for } k = 1, 2 \text{ and } a_1^* a_1 + a_2^* a_2 = r(c^2).$$

Proof. Let $p = r(c^2) \in \mathcal{M}_{\text{proj}}$ be the range projection of c^2. Then $c_1 = c_1 p = p c_1$ and $c = cp = pc$; indeed, by assumption and Lemma 15

$$0 \leq (1-p) c_1^2 (1-p) \leq (1-p) c^2 (1-p) = c^2 - pc^2 - c^2 p + pc^2 p = 0,$$

which gives $(1-p)c_1 = 0$ (the same argument works for c). Consider now the compression $\mathcal{M}_p$ of $\mathcal{M}$ (all operators on pH which are restrictions of operators $pu, u \in \mathcal{M}$) which is a von Neumann algebra of operators on pH. The restrictions of c_1 and c to pH give positive and selfadjoint operators x_1 and y on pH; note that $x_1^2 \leq y^2$, and y^2 by Lemma 16 is injective. Using Lemma 16 again define the contraction $u_1 := x_1 y^{-1}$ on pH which being affiliated to $\mathcal{M}_p$ belongs to $\mathcal{M}_p$. But then $a_1 := i_p u_1 \pi_p$ is a contraction in $\mathcal{M}$ ($\pi_p : H \to pH$ and $i_p : pH \to H$ the canonical mappings) satisfying

$$c_1 = i_p x_1 \pi_p = i_p u_1 y \pi_p = i_p u_1 c p = a_1 c.$$

Finally, if $c = (c_1^2 + c_2^2)^{1/2}$, then with u_1, u_2, a_1, a_2 as above

$$u_1^* u_1 + u_2^* u_2 = \sum_{k=1}^{2} (y^{-1})^* x_k^* x_k y^{-1} = \sum_{k=1}^{2} y^{-1} x_k^2 y^{-1} = y^{-1} y^2 y^{-1} = 1$$

(use that $(y^{-1})^* x_k^* \subset (x_k y^{-1})^*$) which clearly gives

$$a_1^* a_1 + a_2^* a_2 = i_p (u_1^2 + u_2^2) \pi_p = i_p \pi_p = p,$$

the desired equality.   □

The following result is absolutely crucial – though many aspects of its proof are standard we give it with all details.

Proposition 3. *Let $E(\mathcal{M}, \tau)$ be a symmetric space of operators. Then*

(1) $E(\mathcal{M}, \tau)[\ell_\infty(I)]$ is a Banach space.
(2) $E(\mathcal{M}, \tau)[\ell_\infty^c(I)]$ and $E(\mathcal{M}, \tau)[\ell_\infty^r(I)]$ are Banach spaces provided the Banach function space E is 2-convex with constant 1.
(3) $E(\mathcal{M}, \tau)[\ell_\infty^c(I)]$, $E(\mathcal{M}, \tau)[\ell_\infty^r(I)]$, and $E(\mathcal{M}, \tau)[\ell_\infty^{r+c}(I)]$ are quasi Banach spaces.

See the remark after Example 2 which tells that it is not possible to avoid the convexity assumption in statement (2).

Proof. We first prove (1); clearly, we may assume that $I = \mathbb{N}$. Let us first check that the norm $\| \cdot \|_{E[\ell_\infty]}$ satisfies the triangle inequality. Take two sequences $(x_1(n))$ and $(x_2(n))$ in $E(\mathcal{M})$ which for $k = 1, 2$ allow uniform factorizations $x_k(n) = a_k y_k(n) b_k$ with $\|a_k\|_{E^{1/2}(\mathcal{M})} = \|b_k\|_{E^{1/2}(\mathcal{M})} = \|x_k\|_{E[\ell_\infty]}^{1/2}$ and $\sup_n \|y_k(n)\|_\infty \leq 1$. By taking polar decompositions $a_k = |a_k^*| u_k$ and $b_k = v_k |b_k|$ and substituting $y_k(n)$ by $u_k y_k(n) v_k$, we may and will assume that the a_k's and b_k's are positive. Define the operators

$$a := (a_1^2 + a_2^2)^{1/2} \in E^{1/2}(M, \tau) \text{ and } b := (b_1^2 + b_2^2)^{1/2} \in E^{1/2}(M, \tau);$$

clearly,

$$\|a\|_{E^{1/2}(\mathcal{M})} = \|a_1^2 + a_2^2\|_{E(\mathcal{M})}^{1/2}$$

$$\leq (\|a_1^2\|_{E(\mathcal{M})} + \|a_2^2\|_{E(\mathcal{M})})^{1/2} = (\|x_1\|_{E[\ell_\infty]} + \|x_2\|_{E[\ell_\infty]})^{1/2},$$

and similarly $\|b\|_{E^{1/2}(\mathcal{M})} \leq (\|x_1\|_{E[\ell_\infty]} + \|x_2\|_{E[\ell_\infty]})^{1/2}$. By Lemma 17 there are positive contractions $u_k \in \mathcal{M}$ and $v_k \in \mathcal{M}$ such that $a_k = a u_k$, $b_k = v_k b$ and $u_1 u_1^* + u_2 u_2^* = \mathrm{r}(a^2)$, $v_1^* v_1 + v_2^* v_2 = \mathrm{r}(b^2)$. Obviously,

$$x_1 + x_2 = a u_1 y_1(\cdot) v_1 b + a u_2 y_2(\cdot) v_2 b = a\big(u_1 y_1(\cdot) v_1 + u_2 y_2(\cdot) v_2\big) b.$$

Now define in $\mathcal{M}$ the sequence

$$y := u_1 y_1(\cdot) v_1 + u_2 y_2(\cdot) v_2.$$

In $M_2(\mathcal{M})$ (all 2×2 matrices with entries from $\mathcal{M}$) we have the equality

$$\begin{pmatrix} y(n) & 0 \\ 0 & 0 \end{pmatrix} = \begin{pmatrix} u_1 & u_2 \\ 0 & 0 \end{pmatrix} \begin{pmatrix} y_1(n) & 0 \\ 0 & y_2(n) \end{pmatrix} \begin{pmatrix} v_1 & 0 \\ v_2 & 0 \end{pmatrix},$$

and since moreover all three matrices on the right side define contractions on $\ell_2^2(H)$ (have norm ≤ 1 in $M_2(\mathcal{M})$) we get that

$$\sup_n \|y(n)\|_\infty = \sup_n \left\| \begin{pmatrix} y(n) & 0 \\ 0 & 0 \end{pmatrix} \right\|_{M_2(\mathcal{M})}$$

$$\leq \left\| \begin{pmatrix} u_1 & u_2 \\ 0 & 0 \end{pmatrix} \right\|_{M_2(\mathcal{M})} \left\| \begin{pmatrix} y_1(n) & 0 \\ 0 & y_2(n) \end{pmatrix} \right\|_{M_2(\mathcal{M})} \left\| \begin{pmatrix} v_1 & 0 \\ v_2 & 0 \end{pmatrix} \right\|_{M_2(\mathcal{M})} \leq 1;$$

indeed, for the third (and analogously for the first) matrix note that for every $\xi = (\xi_1, \xi_2) \in \ell_2^2(H)$ this follows from

$$\left\| \begin{pmatrix} v_1 & 0 \\ v_2 & 0 \end{pmatrix} \begin{pmatrix} \xi_1 \\ \xi_2 \end{pmatrix} \right\| = (\|v_1\xi_1\|^2 + \|v_2\xi_1\|^2)^{1/2}$$

$$= ((v_1^*v_1 + v_2^*v_2)\xi_1|\xi_1)^{1/2} = (r(b^2)\xi_1|\xi_1)^{1/2} \le \|\xi_1\| \le \|\xi\|.$$

All together we as desired conclude that

$$\|x_1 + x_2\|_{E[\ell_\infty]} = \|ay(\cdot)b\|_{E[\ell_\infty]}$$
$$\le \|a\|_{E^{1/2}(\mathcal{M})} \sup_n \|y(n)\|_\infty \|b\|_{E^{1/2}(\mathcal{M})} \le \|x_1\|_{E[\ell_\infty]} + \|x_2\|_{E[\ell_\infty]}.$$

For the proof of the completeness let (x_k) be a Cauchy sequence in $E(\mathcal{M})[\ell_\infty]$; without loss of generality we may assume that for all k

$$\|x_k - x_{k+1}\|_{E[\ell_\infty]} \le 2^{-3k}/2.$$

Define for each N the sequences

$$u_N := \sum_{k=N}^\infty x_{k+1}(\cdot) - x_k(\cdot)$$

in $E(\mathcal{M})$; we will show that all of them belong to $E(\mathcal{M})[\ell_\infty]$. Choose factorizations

$$x_{k+1}(\cdot) - x_k(\cdot) = a_k y_k(\cdot) b_k$$

with $\|a_k\|_{E^{1/2}(\mathcal{M})} \le 2^{-k}$, $\|b_k\|_{E^{1/2}(\mathcal{M})} \le 2^{-k}$, $\sup_n \|y_k(n)\|_\infty \le 2^{-k}$, and assume again without loss of generality that the a_k and b_k are positive. Obviously,

$$\sum_1^\infty a_k^2 \in E(\mathcal{M}) \text{ and } \sum_1^\infty b_k^2 \in E(\mathcal{M})$$

with norms ≤ 1; indeed,

$$\sum_1^\infty \|a_k^2\|_{E(\mathcal{M})} = \sum_1^\infty \|a_k\|_{E^{1/2}(\mathcal{M})}^2 \le \sum_1^\infty 2^{-2k} \le 1$$

$$\sum_1^\infty \|b_k^2\|_{E(\mathcal{M})} = \sum_1^\infty \|b_k\|_{E^{1/2}(\mathcal{M})}^2 \le \sum_1^\infty 2^{-2k} \le 1.$$

(3.7)

Define

$$a := \left(\sum_1^\infty a_k^2 \right)^{1/2} \in E^{1/2}(\mathcal{M}) \text{ and } b := \left(\sum_1^\infty b_k^2 \right)^{1/2} \in E^{1/2}(\mathcal{M}).$$

Both operators have norm ≤ 1, and according to Lemma 17 there are contractions u_k and v_k in $\mathcal{M}$ which satisfy $au_k = a_k$ and $v_k b = b_k$. Define for each N the sequence

$$v_N := \sum_{k=N}^\infty u_k y_k(\cdot) v_k$$

in $\mathcal{M}$, and note that $\sup_n \|v_N(n)\|_\infty \leq 2^{-N+1}$:

$$\sum_{k=N}^\infty \|u_k y_k(n) v_k\|_\infty \leq \sum_{k=N}^\infty \sup_m \|y_k(m)\|_\infty \leq \sum_{k=N}^\infty 2^{-k} \leq 2^{-N+1}.$$

Then

$$u_N = \sum_{k=N}^\infty a_k y_k(\cdot) b_k = \sum_{k=N}^\infty a u_k y_k(\cdot) v_k b = a v_N(\cdot) b,$$

so that $u_N \in E(\mathcal{M})[\ell_\infty]$ and moreover

$$\|u_N\|_{E[\ell_\infty]} = \|a v_N(\cdot) b\|_{E[\ell_\infty]} \leq \|a\|_{E^{1/2}(\mathcal{M})} \sup_n \|v_N(n)\|_\infty \|b\|_{E^{1/2}(\mathcal{M})} \leq 2^{-N+1}.$$

Finally, we obtain that

$$\left\| \sum_{k=1}^{N-1} (x_{k+1} - x_k) - u_1 \right\|_{E[\ell_\infty]} = \|u_N\|_{E[\ell_\infty]} \leq 2^{-N+1},$$

which clearly shows that in $E(\mathcal{M})[\ell_\infty]$

$$x_N = \sum_{k=1}^{N-1} (x_{k+1} - x_k) + x_1 \longrightarrow u_1 + x_1,$$

the conclusion.

The proof of (2) is similar; we only do the column case: Assume that E is 2-convex with 2-convexity constant 1. We show that $E(\mathcal{M})[\ell_\infty^c]$ is a Banach space, and concentrate on the triangle inequality. Note first the following obvious reformulation of the norm in $E(\mathcal{M})[\ell_\infty^c]$: For $(x(n)) \in E(\mathcal{M})[\ell_\infty^c]$ we have

$$\|(x(n))\|_{E[\ell_\infty^c]} = \inf \|a\|_\infty \sup_n \|y(n)\|_\infty \|b\|_{E^{1/2}(\mathcal{M})},$$

the infimum now taken over all possible factorizations

$$x(n) = a y(n) b, \ n \in \mathbb{N}$$

with $a \in \mathcal{M}$, $b \in E(\mathcal{M})$ and $y = (y(n))$ a bounded sequence in $\mathcal{M}$. Hence this time it suffices to consider two sequences $(x_1(n))$ and $(x_2(n))$ in $E(\mathcal{M})$ which have uniform factorizations $x_k(n) = a_k y_k(n) b_k$ with

$$\|a_k\|_\infty = \|b_k\|_{E(\mathcal{M})} = \|x_k\|_{E[\ell_\infty^c]}^{1/2} \quad \text{and} \quad \sup_n \|y_k(n)\|_\infty \le 1.$$

As above, we can assume that all a_k and b_k are positive, and define the two operators

$$a := (a_1^2 + a_2^2)^{1/2} \in \mathcal{M} \quad \text{and} \quad b := (b_1^2 + b_2^2)^{1/2} \in E(\mathcal{M}).$$

Since $a \in \mathcal{M}$ is selfadjoint, we have that

$$\|a\|_\infty^2 = \|a^2\|_\infty = \|a_1^2 + a_2^2\|_\infty \le \|a_1^2\|_\infty + \|a_2^2\|_\infty,$$

and therefore

$$\|a\|_\infty \le \left(\|a_1\|_\infty^2 + \|a_2\|_\infty^2\right)^{1/2} = \left(\|x_1\|_{E[\ell_\infty^c]} + \|x_2\|_{E[\ell_\infty^c]}\right)^{1/2}.$$

Moreover, E is 2-convex with constant 1, hence by the triangle inequality in $E^2(\mathcal{M})$ we have

$$\|b\|_{E(\mathcal{M})} = \|b_1^2 + b_2^2\|_{E^2(\mathcal{M})}^{1/2}$$

$$\le \left(\|b_1^2\|_{E^2(\mathcal{M})} + \|b_2^2\|_{E^2(\mathcal{M})}\right)^{1/2} \tag{3.8}$$

$$= \left(\|b_1\|_{E(\mathcal{M})}^2 + \|b_2\|_{E(\mathcal{M})}^2\right)^{1/2} = \left(\|x_1\|_{E[\ell_\infty^c]} + \|x_2\|_{E[\ell_\infty^c]}\right)^{1/2}.$$

We now proceed as above; we choose appropriate contractions $u_k, v_k \in \mathcal{M}$ with $a_k = a u_k$ and $b_k = v_k b$ such that the representation

$$x_1 + x_2 = a(u_1 y_1(\cdot) v_1 + u_2 y_2(\cdot) v_2) b$$

gives

$$\|x_1 + x_2\|_{E[\ell_\infty^c]} \le \|x_1\|_{E[\ell_\infty^c]} + \|x_2\|_{E[\ell_\infty^c]},$$

the desired triangle inequality in the column case. Completeness: Take a Cauchy sequence (x_k) in $E(\mathcal{M})[\ell_\infty^c]$ with $\|x_k - x_{k+1}\|_{E[\ell_\infty^c]} \le 2^{-2k}/2$. We again for each N define the sequences $u_N := \sum_{k=N}^\infty x_{k+1}(\cdot) - x_k(\cdot) \in E(\mathcal{M})$, and show that all belong to $E(\mathcal{M})[\ell_\infty^c]$. Choose factorizations $x_{k+1}(\cdot) - x_k(\cdot) = y_k(\cdot) b_k$ with $\|b_k\|_{E(\mathcal{M})} \le 2^{-k}$, $\sup_n \|y_k(n)\|_\infty \le 2^{-k}$, and assume again without loss of generality that the b_k are positive. Since

$$\sum_1^\infty \|b_k^2\|_{E^2} = \sum_1^\infty \|b_k\|_{E(\mathcal{M})}^2 \le \sum_1^\infty 2^{-2k} \le 1, \tag{3.9}$$

we see that $\sum_1^\infty b_k^2 \in E^2(\mathcal{M})$ with norm ≤ 1. Define $b = \left(\sum_1^\infty b_k^2\right)^{1/2} \in E(\mathcal{M})$ and choose contractions $v_k \in \mathcal{M}$ with $v_k b = b_k$. Then we can again define $v_N = \sum_{k=N}^\infty y_k(\cdot) v_k \in \mathcal{M}$, and get that $u_N = v_N(\cdot)b$ and $\|u_N\|_{E[\ell_\infty^c]} \leq 2^{-N+1}$. Now the proof finishes exactly as above.

Let us finally sketch the proof of (3). We show that $E(\mathcal{M})[\ell_\infty^c]$ is always a quasi Banach space: Now the symmetric Banach function space E^2 is only a quasi Banach lattice. Nevertheless, we can define

$$E^2(\mathcal{M}) := \left\{x \in L_0(\mathcal{M}, \tau) \mid |x|^{1/2} \in E(\mathcal{M})\right\}, \; \|x\|_{E^2(\mathcal{M})} = \||x|^{1/2}\|_{E(\mathcal{M})}^2$$

exactly as above, and it was shown in [96] that $E^2(\mathcal{M})$ is a quasi Banach space satisfying

$$\|z_1 + z_2\|_{E^2(\mathcal{M})} \leq \left(\|z_1\|_{E^2(\mathcal{M})}^{1/2} + \|z_2\|_{E^2(\mathcal{M})}^{1/2}\right)^2 \text{ for all } z_1, z_2 \in E^2(\mathcal{M}). \quad (3.10)$$

Modifying (3.8) we then see that for all $x_1, x_2 \in E[\ell_\infty^c]$

$$\|x_1 + x_2\|_{E[\ell_\infty^c]} \leq \sqrt{2}\left(\|x_1\|_{E[\ell_\infty^c]} + \|x_2\|_{E[\ell_\infty^c]}\right).$$

For the proof of the completeness of the quasi normed space $E[\ell_\infty^c]$ we proceed as above – we only have to modify the argument which assures that the series $\sum_1^\infty b_k^2$ converges in the quasi Banach space $E^2(\mathcal{M})$. But this holds since by (3.10) we have that

$$\left\|\sum_{k=N}^M b_k^2\right\|_{E^2} \leq \left(\sum_{k=N}^M \|b_k^2\|_{E^2}^{1/2}\right)^2 = \left(\sum_{k=N}^M \|b_k\|_E\right)^2 \leq \left(\sum_{k=N}^M 2^{-k}\right)^2.$$

Finally, since $E(\mathcal{M})[\ell_\infty^{r+c}]$ is nothing else than the sum of the two quasi Banach spaces $E(\mathcal{M})[\ell_\infty^r]$ and $E(\mathcal{M})[\ell_\infty^c]$ the proof of Proposition 3 is complete.   □

As in the commutative case, we later want to derive from maximal inequalities results on almost uniform convergence. We follow the lines of Sect. 2.1.2 and define

$$E(\mathcal{M})[c_0] = E(\mathcal{M}, \tau)[c_0(I)]$$

as a subset of $E(\mathcal{M}, \tau)[\ell_\infty(I)]$ (in fact it turns out to be a subspace); by definition it consists of all families $(x_n)_{n \in I} \in E(\mathcal{M}, \tau)[\ell_\infty(I)]$ for which there is a factorization $x_n = a y_n b$, where again $a, b \in E^{1/2}(\mathcal{M}, \tau)$, but now the sequence (y_n) is a zero sequence in $\mathcal{M}$. Clearly, we endow this space with the quasi norm defined by the infimum taken with respect to all possible representations of such type. By now the definitions of:

- $E(\mathcal{M})[c_0^c]$, the column variant
- $E(\mathcal{M})[c_0^r]$, the row variant
- $E(\mathcal{M})[c_0^{r+c}]$, the row+column variant

as subspaces of $E(\mathcal{M})[\ell_\infty^c]$, $E(\mathcal{M})[\ell_\infty^r]$, and $E(\mathcal{M})[\ell_\infty^{r+c}]$, respectively, are obvious.

Proposition 4. $E(\mathcal{M}, \tau)[c_0(I)]$ *is a Banach space which embeds isometrically into* $E(\mathcal{M}, \tau)[\ell_\infty(I)]$; *the same result holds in the column, row and row+column case.*

Proof. We prove the symmetric case. That $E(\mathcal{M})[c_0]$ is a Banach space follows by an easy inspection of the proof of Proposition 3. It remains to show that $E(\mathcal{M})[c_0]$ is an isometric subspace of $E(\mathcal{M})[\ell_\infty]$. Clearly, for a given sequence $(x_n) \in E(\mathcal{M})[c_0]$ we have

$$\|(x_n)\|_{E(\mathcal{M})[\ell_\infty]} \leq \|(x_n)\|_{E(\mathcal{M})[c_0]}.$$

Assume conversely that $(x_n) \in E(\mathcal{M})[\ell_\infty]$ with $\|(x_n)\|_{E(\mathcal{M})[\ell_\infty]} < 1$. Choose a factorization

$$x_n = a y_n b$$

with $\|a\|_{E^{1/2}(\mathcal{M})} = \|b\|_{E^{1/2}(\mathcal{M})} < 1$ and $\sup_n \|y_n\|_\infty \leq 1$. We want to show that

$$\|(x_n)\|_{E(\mathcal{M})[c_0]} < 1,$$

and hence take an arbitrary factorization

$$x_n = c z_n d$$

with $c, d \in E^{1/2}(\mathcal{M})$ and $\lim_n \|z_n\|_\infty = 0$. Choose $\varepsilon > 0$ such that

$$r := (a^*a + \varepsilon c^*c)^{1/2} \in E^{1/2}(\mathcal{M}) \quad \text{and} \quad \|r\|_{E^{1/2}(\mathcal{M})} < 1$$

$$s := (b^*b + \varepsilon d^*d)^{1/2} \in E^{1/2}(\mathcal{M}) \quad \text{and} \quad \|s\|_{E^{1/2}(\mathcal{M})} < 1.$$

Use Lemma 17 to choose operators $u, v, \alpha, \beta \in \mathcal{M}$ such that

$$a = ru \quad \text{with} \quad \|u\|_\infty \leq 1 \quad \text{and} \quad b = vs \quad \text{with} \quad \|v\|_\infty \leq 1$$

as well as

$$c = r\alpha \quad \text{and} \quad d = \beta s.$$

Then $x_n = a y_n b = r u y_n v s$ and $x_n = c z_n d = r \alpha z_n \beta s$. Let now e and f be the range projection of r and s, respectively. Then

$$e u y_n v f = e \alpha z_n \beta f,$$

hence

$$\sup_n \|e \alpha z_n \beta f\|_\infty \leq 1.$$

On the other hand we have

$$x_n = r e \alpha z_n \beta f s \quad \text{and} \quad \lim_n \|e \alpha z_n \beta f\|_\infty = 0,$$

which finally proves as desired

$$\|(x_n)\|_{E[c_0]} \leq \sup \|e\alpha z_n \beta f\|_\infty \|r\|_{E^{1/2}(\mathcal{M})} \|s\|_{E^{1/2}(\mathcal{M})} < 1,$$

the conclusion. The row and the column case have similar proofs, and the row+column case is then an immediate consequence.   □

3.1.3   Tracial Extensions of Maximizing Matrices

Recall from Theorem 5 that every $(2,2)$-maximizing matrix $A = (a_{jk})$ is even (p,∞)-maximizing which by Definition 1 means that for each unconditionally summable sequence (x_k) in some $L_p(\mu)$ the maximal function $\sup_j |\sum_{k=0}^\infty a_{jk}x_k|$ is p-integrable. In Theorem 13 we even proved that for any unconditionally summable sequences (x_k) in some vector-valued Banach function space $E(\mu,X)$ the maximal function

$$\sup_j \left\| \sum_{k=0}^\infty a_{jk}x_k(\cdot) \right\|_X \in E(\mu).$$

We now show without too much new effort that here vector-valued Banach function spaces $E(\mu,X)$ may be replaced by symmetric spaces $E(\mathcal{M},\tau)$ of operators.

Theorem 17. *Let $A = (a_{jk})$ be a $(2,2)$-maximizing matrix. Then for each unconditionally convergent series $\sum_k x_k$ in $E(\mathcal{M},\tau)$*

$$\left(\sum_{k=0}^\infty a_{jk}x_k \right)_j \in E(\mathcal{M})[\ell_\infty]. \tag{3.11}$$

Moreover, ℓ_∞ can be replaced by ℓ_∞^c provided E is 2-convex.

See Theorem 19 for an improvement whenever E is 2-concave.

Proof. The proof of (3.11) is similar to the proof of Theorem 13 – it suffices to show that

$$\| \mathrm{id}_{E(\mathcal{M})} \otimes A : E(\mathcal{M}) \otimes_\varepsilon \ell_1 \longrightarrow E(\mathcal{M})[\ell_\infty] \| \leq C\, m_{2,2}(A); \tag{3.12}$$

indeed, the following three facts then complete the proof of (3.11): by continuous extension (3.12) implies that

$$\| \mathrm{id}_{E(\mathcal{M})} \otimes A : E(\mathcal{M}) \tilde{\otimes}_\varepsilon \ell_1 \longrightarrow E(\mathcal{M})[\ell_\infty] \| \leq C\, m_{2,2}(A),$$

the completion $E(\mathcal{M}) \tilde{\otimes}_\varepsilon \ell_1$ of the injective tensor product $E(\mathcal{M}) \otimes_\varepsilon \ell_1$ can be identified with all unconditionally convergent series in $E(\mathcal{M})$ (see Sect. 2.1.6 and [6, Sect. 8.1]), and

$$\left(\mathrm{id}_{E(\mathcal{M})} \otimes A \right) \left(\sum_k x_k \otimes e_k \right) = \sum_j \left(\sum_k a_{jk}x_k \right) \otimes e_j.$$

For the proof of (3.12) we again conclude from Grothendieck's théorème fondamental (2.24) and Theorem 4, (1) that $\iota(A) \le K_G \, m_{2,2}(A)$ (K_G Grothendieck's constant), and as a consequence from (2.23) that

$$\| \mathrm{id}_{E(\mathcal{M})} \otimes A : E(\mathcal{M}) \otimes_\varepsilon \ell_1 \longrightarrow E(\mathcal{M}) \otimes_\pi \ell_\infty \| \le \iota(A) \le K_G \, m_{2,2}(A).$$

But

$$\| \mathrm{id} : E(\mathcal{M}) \otimes_\pi \ell_\infty \to E(\mathcal{M})[\ell_\infty] \| \le 1;$$

indeed, for $\xi \in \ell_\infty$ and $x \in E(\mathcal{M})$ with polar decomposition $x = u|x|$ we have

$$\left\| (x\xi_i) \right\|_{E[\ell_\infty]} = \left\| (u|x|\xi_i) \right\|_{E[\ell_\infty]}$$

$$= \left\| \left(u|x|^{1/2} \xi_i |x|^{1/2} \right) \right\|_{E[\ell_\infty]}$$

$$\le \left\| u|x|^{1/2} \right\|_{E^{1/2}(\mathcal{M})} \sup_i |\xi_i| \left\| |x|^{1/2} \right\|_{E^{1/2}(\mathcal{M})}$$

$$\le \left\| |x|^{1/2} \right\|_{E^{1/2}(\mathcal{M})}^2 \sup_i |\xi_i| = \|x\|_{E(\mathcal{M})} \sup_i |\xi_i|.$$

As desired we conclude (3.12). The column case follows in a similar way. □

The preceding result handles unconditionally summable sequences (x_k) in symmetric spaces $E(\mathcal{M}, \tau)$ of operators – within more restrictive classes of sequences (x_k) and noncommutative L_p-spaces it can be extended considerably. Recall from Definition 1 that a matrix $A = (a_{jk})$ is (p,q)-maximizing whenever for each $\alpha \in \ell_q$ and each weakly q'-summable sequence (x_k) in an arbitrary $L_p(\mu)$ we have

$$\left(\sum_{k=0}^\infty a_{jk} \alpha_k x_k \right)_{j=0}^\infty \in L_p(\mu)[\ell_\infty].$$

The following result shows that here the commutative L_p-spaces can be substituted by any noncommutative $L_p(\mathcal{M}, \tau)$. As in the preceding theorem we carefully have to distinguish between symmetric and left/righthanded factorizations of the linear means $\sum_{k=0}^\infty a_{jk} \alpha_k x_k$.

Theorem 18. *Let $A = (a_{jk})$ be a (p,q)-maximizing matrix. Then for each $\alpha \in \ell_q$ and each weakly q'-summable sequence (x_k) in $L_p(\mathcal{M}, \tau)$*

$$\left(\sum_{k=0}^\infty a_{jk} \alpha_k x_k \right)_j \in L_p(\mathcal{M})[\ell_\infty].$$

Moreover, ℓ_∞ can be replaced by ℓ_∞^c provided $p \ge 2$.

The proof of this result is very similar to the one given for Theorem 14. The convexity argument (2.70) used there has to be substituted by an estimate based on complex interpolation of the Banach spaces $L_p(\mathcal{M}, \tau)[\ell_\infty]$. The following important interpolation result is taken from [42, Proposition 2.4] of Junge and Xu (for complex interpolation of Banach spaces see the monograph [4], and for complex interpolation of noncommutatice L_p-spaces see the extensive discussion from [80, Sect. 2]).

Proposition 5. *Let* $1 \le p_0 < p_1 \le \infty$ *and* $0 < \theta < 1$. *Then we have*

$$L_p(\mathcal{M}, \tau)[\ell_\infty] = \Big(L_{p_0}(\mathcal{M}, \tau)[\ell_\infty], L_{p_1}(\mathcal{M}, \tau)[\ell_\infty]\Big)_\theta,$$

where $\frac{1}{p} = \frac{1-\theta}{p_0} + \frac{\theta}{p_1}$. *For* $2 \le p_0 < p_1 \le \infty$ *this interpolation formula also holds in the column case, i.e.* ℓ_∞ *can be replaced by* ℓ_∞^c.

The following result is our substitute for (2.70).

Lemma 18. *Let* $S = (s_{jk})$ *be an* $n \times m$ *matrix and* $1 \le p \le \infty$. *Then for each choice of* $x_0, \ldots, x_n \in L_p(\mathcal{M}, \tau)$

$$\Big\|\Big(\sum_{k=0}^m s_{jk} x_k\Big)_{j=0}^n\Big\|_{L_p[\ell_\infty^n]} \le C \Big(\sum_{k=0}^m \|x_k\|_p^p\Big)^{1/p} \sup_{j=0,\ldots,n} \Big(\sum_{k=0}^m |s_{jk}|^{p'}\Big)^{1/p'},$$

$C > 0$ *some constant independent of* n. *For* $p \ge 2$ *we may replace* ℓ_∞^n *by* $(\ell_\infty^n)^c$.

Note that in terms of tensor products this result reads as follows:

$$\|S \otimes \text{id} : \ell_p^m(L_p(\mathcal{M})) \longrightarrow L_p(\mathcal{M})[\ell_\infty^n]\| \le C \|S : \ell_p^m \longrightarrow \ell_\infty^n\|; \tag{3.13}$$

here $\ell_p^m(L_p(\mathcal{M}))$ as usual stands for the Banach space of all $(m+1)$-tuples $(x_k)_{k=0}^m$ in $L_p(\mathcal{M})$ endowed with the norm $\big(\sum_{k=0}^n \|x_k\|_p^p\big)^{1/p}$ and we identify $L_p(\mathcal{M})[\ell_\infty^n]$ with $\ell_\infty^n \otimes L_p(\mathcal{M})$.

Proof. The proof is based on complex interpolation. We start with the symmetric case. Choose $0 < \theta < 1$ such that $1/p = \theta/1 + (1-\theta)/\infty$. Clearly, for any $S = (s_{jk})$ we have

$$\|S \otimes \text{id} : \ell_1^m(L_1(\mathcal{M})) \longrightarrow L_1(\mathcal{M})[\ell_\infty^n]\|$$
$$\le \|S \otimes \text{id} : \ell_1^m \otimes_\pi L_1(\mathcal{M}) \longrightarrow \ell_\infty^n \otimes_\pi L_1(\mathcal{M})\| = \|S : \ell_1^m \longrightarrow \ell_\infty^n\|,$$

and

$$\|S \otimes \text{id} : \ell_\infty^m(L_\infty(\mathcal{M})) \longrightarrow L_\infty(\mathcal{M})[\ell_\infty^n)]\|$$
$$= \|S \otimes \text{id} : \ell_\infty^m \otimes_\varepsilon \mathcal{M} \longrightarrow \ell_\infty^n \otimes_\varepsilon \mathcal{M}\| = \|S : \ell_\infty^m \longrightarrow \ell_\infty^n\|.$$

From this we can conclude that the two bilinear maps

$$\phi_1 : \ell_1^m(L_1(\mathcal{M})) \times \mathcal{L}(\ell_1^m, \ell_\infty^n) \longrightarrow L_1(\mathcal{M})[\ell_\infty^n]$$

$$\phi_2 : \ell_\infty^m(L_\infty(\mathcal{M})) \times \mathcal{L}(\ell_\infty^m, \ell_\infty^n) \longrightarrow L_\infty(\mathcal{M})[\ell_\infty^n]$$

$$\phi_i((x_k), S) := (S \otimes \mathrm{id})(x_k)$$

satisfy the norm estimates $\|\phi_i\| \le 1$. By a well-known interpolation formula (see e.g. [4, Sect. 5.1]) we have

$$\left(\ell_1^m(L_1(\mathcal{M})), \ell_\infty^m(L_\infty(\mathcal{M}))\right)_\theta = \ell_p^m(L_p(\mathcal{M})),$$

and by the same formula and a standard identification

$$\left(\mathcal{L}(\ell_1^m, \ell_\infty^n), \mathcal{L}(\ell_\infty^m, \ell_\infty^n)\right)_\theta = \left(\ell_\infty^n(\ell_\infty^m), \ell_\infty^n(\ell_1^m)\right)_\theta = \ell_\infty^n(\ell_{p'}^m) = \mathcal{L}(\ell_p^m, \ell_\infty^n).$$

Moreover, we know from Proposition 5

$$\left(L_1(\mathcal{M})[\ell_\infty^n], L_\infty(\mathcal{M})[\ell_\infty^n]\right)_\theta = L_p(\mathcal{M})[\ell_\infty^n]. \qquad (3.14)$$

Hence, by complex bilinear interpolation (see [4, Sect. 4.4]) we conclude that the bilinear mapping

$$\phi : \ell_p^m(L_p(\mathcal{M})) \times \mathcal{L}(\ell_p^m, \ell_\infty^n) \longrightarrow L_p(\mathcal{M})[\ell_\infty^n]$$

$$\phi((x_k), S) := (S \otimes \mathrm{id})(x_k)$$

is bounded with norm ≤ 1, or equivalently for all $S \in \mathcal{L}(\ell_p^m, \ell_\infty^n)$ as desired the estimate (3.13) holds.

The proof of the column case is similar. We assume that $p \ge 2$. Again we have

$$\|S \otimes \mathrm{id} : \ell_\infty^m(L_\infty(\mathcal{M})) \longrightarrow L_\infty(\mathcal{M})[(\ell_\infty^n)^c]\| \le \|S : \ell_\infty^m \longrightarrow \ell_\infty^n\|.$$

Moreover, we check that

$$\|S \otimes \mathrm{id} : \ell_2^m(L_2(\mathcal{M})) \longrightarrow L_2(\mathcal{M})[(\ell_\infty^n)^c]\| \le K_{LG}\|S : \ell_2^m \longrightarrow \ell_\infty^n\|,$$

K_{LG} the little Grothendieck constant; indeed, by the little Grothendieck theorem (2.25) we have

$$\|S' \otimes \mathrm{id} : \ell_1^n \otimes_\varepsilon L_2(\mathcal{M}) \longrightarrow \ell_2^m(L_2(\mathcal{M}))\| = \pi_2(S' : \ell_1^n \longrightarrow \ell_2^m) \le K_{LG}\|S\|,$$

hence by duality

$$\|S \otimes \mathrm{id} : \ell_2^m(L_2(\mathcal{M})) \longrightarrow L_2(\mathcal{M})[(\ell_\infty^n)^c]\|$$
$$\le \|S \otimes \mathrm{id} : \ell_2^m(L_2(\mathcal{M})) \longrightarrow \ell_\infty^n \otimes_\pi L_2(\mathcal{M})\| \le K_{LG}\|S\|.$$

But since for $0 < \theta < 1$ with $\frac{1}{p} = \frac{\theta}{2} + \frac{1-\theta}{\infty}$

$$\left(L_2(\mathcal{M})[(\ell_\infty^n)^c], L_\infty(\mathcal{M})[(\ell_\infty^n)^c]\right)_\theta = L_p(\mathcal{M})[(\ell_\infty^n)^c] \qquad (3.15)$$

(Proposition 5) the proof completes exactly as in the symmetric case.   □

We are ready to give the

Proof (of Theorem 18). The preceding result at hands the proof is very similar to the one given for Theorem 14. Hence – for the sake of completeness – we here only repeat some details. We fix finitely many scalars $\alpha_0, \ldots, \alpha_n$ and show the estimate

$$\left\| \mathrm{id}_{L_p(\mathcal{M})} \otimes A_n D_\alpha : L_p(\mathcal{M}) \otimes_\varepsilon \ell_{q'}^n \longrightarrow L_p(\mathcal{M})[\ell_\infty^n] \right\| \leq C\, m_{p,q}(A) \|\alpha\|_q \qquad (3.16)$$

($C > 0$ some constant independent of n); this in fact proves that for each choice of $x_0, \ldots, x_n \in L_p(\mathcal{M})$ we have

$$\left\| \left(\sum_{k=0}^n a_{jk} \alpha_k x_k \right)_{j=0}^n \right\|_{L_p(\mathcal{M})[\ell_\infty^n]} \leq C\, m_{p,q}(A) \|\alpha\|_q w_{q'}(x_k) ,$$

so that our conclusion follows from a simple analysis of the density argument presented in the proof of Lemma 3. By the general characterization of (p,q)-maximizing matrices from Theorem 3, as well as (2.19) and (2.18) there is a factorization

$$\begin{array}{ccc}
\ell_{q'}^n & \xrightarrow{\;A_n D_\alpha\;} & \ell_\infty^n \\
{\scriptstyle R}\downarrow & & \uparrow{\scriptstyle S} \\
\ell_\infty^m & \xrightarrow[\;D_\mu\;]{} & \ell_p^m
\end{array}$$

with

$$\|R\| \, \|D_\mu\| \, \|S\| \leq (1+\varepsilon)\, m_{p,q}(A) \|\alpha\|_q . \qquad (3.17)$$

Tensorizing gives the commutative diagram

$$\begin{array}{ccc}
L_p(\mathcal{M}) \otimes_\varepsilon \ell_{q'}^n & \xrightarrow{\;\mathrm{id}_{L_p(\mathcal{M})} \otimes A_n D_\alpha\;} & L_p(\mathcal{M})[\ell_\infty^n] \\
{\scriptstyle \mathrm{id}_{L_p(\mathcal{M})} \otimes R}\downarrow & & \uparrow{\scriptstyle S \otimes \mathrm{id}_{L_p(\mathcal{M})}} \\
L_p(\mathcal{M}) \otimes_\varepsilon \ell_\infty^m & \xrightarrow[\;\mathrm{id}_{L_p(\mathcal{M})} \otimes D_\mu\;]{} & \ell_p^m(L_p(\mathcal{M})) \, ;
\end{array}$$

for the definition of the mapping $S \otimes \mathrm{id}_{L_p(\mathcal{M})}$ identify $L_p(\mathcal{M})[\ell_\infty^n]$ again with the tensor product $\ell_\infty^n \otimes L_p(\mathcal{M})$. Now the metric mapping property of ε implies

$$\left\| \mathrm{id}_{L_p(\mathcal{M})} \otimes R \right\| \leq \|R\|,$$

and moreover we obtain from

$$\left\| \mathrm{id}_{L_p(\mathcal{M})} \otimes D_\mu \left(\sum x_k \otimes e_k \right) \right\|_{\ell_p^m(L_p(\mathcal{M}))} = \left(\sum_k \|\mu_k x_k\|_p^p \right)^{1/p}$$

$$\leq \sup_k \|\dot{x}_k\|_p \left(\sum_k |\mu_k|^p \right)^{1/p}$$

that

$$\left\| \mathrm{id}_{L_p(\mathcal{M})} \otimes D_\mu \right\| \leq \|D_\mu\|.$$

Finally, Lemma 18 assures that

$$\left\| S \otimes \mathrm{id}_{L_p(\mathcal{M})} \right\| \leq C,$$

$C > 0$ a uniform constant. All together we obtain with (3.17) that

$$\left\| \mathrm{id}_{L_p(\mathcal{M})} \otimes A_n D_\alpha \right\| \leq \|R\| \, \|D_\mu\| \, \|S\| \leq (1+\varepsilon) C \, m_{p,q}(A) \|\alpha\|_q,$$

our conclusion (3.16).   □

3.1.4 The Row+Column Maximal Theorem

Let us show that the last statement in the preceding Theorem 17 is in general false whenever E is not 2-convex. Recall from (2.34) that

$$a_{jk} := \begin{cases} \dfrac{1}{\log k} & k \leq j \\ 0 & k > j \end{cases}$$

is our central example of a $(2,2)$-maximizing matrix, and that S_p denotes the Schatten p-class.

Example 2. Let $1 \leq p < 2$. Then there is an unconditionally convergent series $\sum_k x_k$ in S_p such that

$$\left(\sum_{k=0}^{j} \frac{x_k}{\log k} \right)_j \notin S_p[\ell_\infty^c];$$

in other words, there is no uniform factorization

$$\sum_{k=0}^{j} \frac{x_k}{\log k} = z_j a \quad \text{with} \quad a \in S_p, \ \sup_j \|z_j\|_\infty < \infty.$$

Of course, a similar counterexample holds in the row case $S_p[\ell_\infty^r]$.

This example is taken from [7, Example 4.4], and shows also that $S_p[\ell_\infty^c]$ and $S_p[\ell_\infty^r]$ for $1 \le p < 2$ are no Banach spaces; if yes, then the proof of Theorem 17 would show that in fact $\left(\sum_{k=0}^{j} \frac{x_k}{\log k} \right)_j \in S_p[\ell_\infty^c]$ for every unconditionally summable sequence (x_k) of operators in S_p (in contradiction to the example).

Proof. We do the proof in the row case, i.e. we assume that for every unconditionally convergent series $\sum_k x_k$ we have

$$\left(\sum_{k=0}^{j} x_k \right)_j \in S_p[\ell_\infty^r].$$

Since $S_p[\ell_\infty^r]$ is a quasi Banach space (see Proposition 3), a closed graph argument assures a constant $C \ge 1$ such that for each choice of finitely many $x_0, \dots, x_N \in S_p$

$$\left\| \left(\sum_{k=0}^{j} \frac{x_k}{\log k} \right)_{j \le N} \right\|_{S_p[\ell_\infty^r]} \le C \sup_{\|x'\|_{S_{p'}} \le 1} \sum_{k=0}^{N} |x'(x_k)|. \tag{3.18}$$

For N and $0 \le k \le N$ we put $x_k := (N+1)^{-1/2} e_{k1} \in S_p$, and an evaluation of both sides of this inequality for these vectors will lead to a contradiction. For the right side we easily see that

$$\sup_{\|x'\|_{S_{p'}} \le 1} \sum_{k=0}^{N} |x'(x_k)| = \sup_{\|\mu\|_{\ell_\infty^N} \le 1} \left\| \sum_{k=0}^{N} \mu_k x_k \right\|_{S_p} = (N+1)^{-1/2} \sup_{\|\mu\|_{\ell_\infty^N} \le 1} \|\mu\|_{\ell_2^N} = 1.$$

For the left side of (3.18) define for $0 \le j \le N$

$$y_j := \sum_{k=0}^{j} \frac{x_k}{\log k} = \sum_{k=0}^{j} \lambda_k e_{k1} \quad \text{with} \quad \lambda_k := \frac{1}{(N+1)^{1/2} \log k}.$$

Hence by (3.18) there is a uniform factorization

$$y_j = c z_j, \quad 0 \le j \le N \tag{3.19}$$

$$c \in S_p \quad \text{with} \quad \|c\|_{S_p} \le 1$$

$$z_0, \dots, z_N \in \mathscr{L}(\ell_2) \quad \text{with} \quad \sup_j \|z_j\|_\infty \le C+1.$$

Define now

$$s_N : \ell_2^N \longrightarrow \ell_2^N, \quad s_N(\xi) := \left(\sum_{k=0}^{j} \lambda_k \xi_k \right)_{j \leq N}$$

and note that s_N^* is given by the matrix

$$\begin{pmatrix} \lambda_0 & \lambda_0 & . & \lambda_0 \\ 0 & \lambda_1 & . & . \\ . & . & . & . \\ 0 & . & . & \lambda_N \end{pmatrix}.$$

Then, if π_N stands for the canonical projection from ℓ_2 onto ℓ_2^N, we have

$$s_N^* e_j = \pi_N y_j e_1, \quad 0 \leq j \leq N, \tag{3.20}$$

and moreover

$$\|(s_N^*)^{-1} : \ell_2^N \longrightarrow \ell_2^N\| \leq 2 \max_{0 \leq j \leq N} |\lambda_j^{-1}|, \tag{3.21}$$

which follows from the factorization

$$(s_N^*)^{-1} = \begin{pmatrix} 1 & -1 & 0 & . & 0 \\ 0 & . & . & . & . \\ . & . & . & . & 0 \\ . & . & . & . & -1 \\ 0 & . & . & 0 & 1 \end{pmatrix} \begin{pmatrix} \lambda_0^{-1} & 0 & . & . & 0 \\ 0 & . & . & . & 0 \\ . & . & . & . & . \\ . & . & . & . & 0 \\ 0 & . & . & 0 & \lambda_N^{-1} \end{pmatrix}.$$

Then we conclude from (3.20) and (3.19) that

$$s_N^* e_j = \pi_N c z_j e_1,$$

hence for $u_N := (s_N^*)^{-1} \pi_N c \in \mathcal{L}(\ell_2, \ell_2^N)$ we obtain

$$e_j = u_N(z_j e_1), \quad 0 \leq j \leq N. \tag{3.22}$$

In particular, the vectors $z_j e_1 \in \ell_2$ are linearly independent, and as a consequence $E_N := \mathrm{span}\{z_j e_1 \mid 0 \leq j \leq N\} \subset \ell_2$ has dimension $N+1$ and the restriction v_N of u_N to E_N is invertible. Then, it follows from (3.22) and (3.19) that

$$\|v_N^{-1} : \ell_2^N \longrightarrow E_N\|_{S_2^{N+1}} = \left(\sum_{j=0}^{N} \|v_N^{-1}(e_j)\|_2^2 \right)^{1/2}$$

$$\leq (N+1)^{1/2} \sup_{0 \leq j \leq N} \|v_N^{-1} e_j\|_2$$

$$= (N+1)^{1/2} \sup_{0 \leq j \leq N} \|z_j e_1\|_2 \leq (N+1)^{1/2}(C+1)$$

(here and in the following S_p^{N+1} as usual denotes the Schatten p-class on ℓ_2^N; recall that in these notes ℓ_2^N by our convention from the preliminaries and Sect. 2.1.2 is $N+1$-dimensional), and moreover by (3.19) and (3.21)

$$\|v_N\|_{S_p^{N+1}} = \|(s_N^*)^{-1}\pi_N c_{|E_N}\|_{S_p^{N+1}}$$

$$\leq \|(s_N^*)^{-1}\|\,\|c\|_{S_p} \leq 2(N+1)^{1/2}\log(N+1).$$

Altogether we see that for $1/r = 1/p + 1/2$ by the generalized Hölder inequality

$$(N+1)^{1/r} = \|\mathrm{id}_{E_N}\|_{S_r^{N+1}} \leq \|v_N\|_{S_p^{N+1}}\,\|v_N^{-1}\|_{S_2^{N+1}} \leq 2(C+1)(N+1)\log(N+1),$$

contradicting the assumption $1 \leq p < 2$. $\square$

Motivated by the preceding example we now prove the so called row+column maximal theorem – it complements Theorem 17 within 2-concave symmetric operator spaces.

Theorem 19. *Let $E(\mathcal{M}, \tau)$ be a symmetric space of operators where E is 2-concave such that either $E^\times$ is ordercontinuous or $E = L_1$. Then each unconditionally summable series $\sum_k x_k$ in $E(\mathcal{M})$ splits into a sum*

$$\sum_k x_k = \sum_k w_k^r + \sum_k w_k^c$$

of two unconditionally convergent series such that for each $(2,2)$-maximizing matrix $A = (a_{jk})$ we have

$$\left(\sum_{k=0}^\infty a_{jk}w_k^r\right)_j \in E(\mathcal{M})[\ell_\infty^r] \quad \text{and} \quad \left(\sum_{k=0}^\infty a_{jk}w_k^c\right)_j \in E(\mathcal{M})[\ell_\infty^c].$$

Note that this result in particular shows that in the above situation for each unconditionally convergent series $\sum_k x_k$ in $E(\mathcal{M})$ and for each $(2,2)$-maximizing matrix $A = (a_{jk})$, we have

$$\left(\sum_{k=0}^\infty a_{jk}x_k\right)_j \in E(\mathcal{M})[\ell_\infty^{r+c}]. \tag{3.23}$$

The proof is based on a splitting theorem for unconditionally convergent series in 2-concave symmetric operator spaces which itself is a sort of reformulation of an important noncommmutative analogue of the little Grothendieck theorem due to Lust-Piquard and Xu [56, Theorem 1.1] (for an earlier slightly weaker version of this result see [55], and for Grothendieck's original theorem (2.25)): For any operator $u : E(\mathcal{M}, \tau) \longrightarrow H$, where H is a Hilbert space and the Banach function space E is

2-convex, there is a positive norm one functional $\varphi \in E^2(\mathcal{M}, \tau)'$ such that for each $x \in E(\mathcal{M}, \tau)$ we have

$$\|ux\| \le C\|u\|\varphi(xx^* + x^*x)^{1/2}, \qquad (3.24)$$

$C > 0$ some constant.

Lemma 19. *Let $E(\mathcal{M}, \tau)$ be a symmetric space of operators where E is 2-concave such that either $E^\times$ is ordercontinuous or $E = L_1$. Then for each unconditionally summable sequence (x_n) in $E(\mathcal{M})$ there are two unconditionally summable sequences (u_n) and (v_n) in $L_2(\mathcal{M})$ as well as some $a \in E^{\times 2 \times 1/2}(\mathcal{M})$ such that for all n we have*

$$x_n = au_n + v_n a.$$

As a by product the proof shows that

$$M^a : L_2(\mathcal{M}) \oplus_2 L_2(\mathcal{M}) \longrightarrow E(\mathcal{M}), \; M^a(u,v) := au + va \qquad (3.25)$$

is well-defined with norm $\le \sqrt{2}\|a\|_{E^{\times 2 \times 1/2}}$, a fact which will be useful later.

Proof. Define the operator

$$u : c_0 \longrightarrow E(\mathcal{M}), \; u(e_k) := x_k.$$

Since E by assumption is a 2-concave Banach lattice, we may assume without loss of generality that $M_{(2)}(E) = 1$. Moreover, it is well-known that the notions of 2-concavity and cotype 2 are equivalent (see e.g. [9, Sect. 16.9]) which implies that $E(\mathcal{M})$ has cotype 2 (see [18, Theorem 4] and [95]). But then there is a factorization

$$\begin{array}{ccc} c_0 & \xrightarrow{\;\;u\;\;} & E(\mathcal{M}) \\ & \searrow{\scriptstyle v} \quad \uparrow{\scriptstyle w} & \\ & H & \end{array} \qquad (3.26)$$

where H is some Hilbert space and v, w are bounded operators (see e.g [6, Sect. 31.4] or [77, Sect. 4.1]).

Let us first consider the case in which $E^\times$ is ordercontinuous. Note that also E, being 2-concave, is ordercontinuous, hence the dual operator w' by (3.4) defines an operator from $E(\mathcal{M})' = E(\mathcal{M})^\times = E^\times(\mathcal{M})$ into H. Moreover from (3.24) we conclude that there is some positive norm one functional $\varphi \in E^{\times 2}(\mathcal{M})'$ such that for all $x \in E^\times(\mathcal{M})$

$$\|w'x\| \le C\|w'\|\varphi(x^*x + xx^*)^{1/2},$$

$C > 0$ some constant. Since $E^{\times 2}$ by assumption is ordercontinuous, we also know that $E^{\times 2}(\mathcal{M})' = E^{\times 2\times}(\mathcal{M})$. Hence the preceding inequality yields some positive $d \in E^{\times 2\times}(\mathcal{M})$ such that for all $x \in E^{\times}(\mathcal{M})$ we have

$$\|w'x\| \leq C\|w\| \tau(d(x^*x + xx^*))^{1/2}.$$

Now since $0 \leq d \in E^{\times 2\times}(\mathcal{M})$ and $0 \leq x^*x + xx^* \in E^{\times 2}(\mathcal{M})$ we conclude from (3.6) and with $a := d^{1/2} \in E^{\times 2\times 1/2}(\mathcal{M})$ that for all $x \in E^{\times}(\mathcal{M})$

$$\tau(d(x^*x + xx^*)) = \tau(d^{1/2}(x^*x + xx^*)d^{1/2}) = \tau(|xa|^2 + |ax|^2), \tag{3.27}$$

therefore

$$\|w'x\| \leq C\|w\| \tau(|xa|^2 + |ax|^2)^{1/2}. \tag{3.28}$$

Define the multiplication operator

$$M_a : E^{\times}(\mathcal{M}) \longrightarrow L_2(\mathcal{M}) \oplus_2 L_2(\mathcal{M}), \; M_a(x) := ax + xa,$$

which is well-defined and by (3.27) satisfies the following norm estimate:

$$\sup_{\|x\|_{E^{\times}(\mathcal{M})} \leq 1} \|M_a(x)\|_{L_2(\mathcal{M}) \oplus_2 L_2(\mathcal{M})} = \sup_{\|x\|_{E^{\times}(\mathcal{M})} \leq 1} (\tau(|ax|^2) + \tau(|xa|^2))^{1/2}$$

$$= \sup_{\|x\|_{E^{\times}(\mathcal{M})} \leq 1} \tau(d(x^*x + xx^*))^{1/2} \tag{3.29}$$

$$\leq \sqrt{2} \, \|d\|_{E^{\times 2\times}}^{1/2}.$$

By duality we show that the operator

$$M^a : L_2(\mathcal{M}) \oplus_2 L_2(\mathcal{M}, \tau) \longrightarrow E(\mathcal{M}), \; M^a(u, v) := au + va \tag{3.30}$$

is well-defined: check first that for $v \in L_2(\mathcal{M})$ and $x \in E^{\times}(\mathcal{M})$

$$\tau(axv) = \tau(xva); \tag{3.31}$$

this follows from (3.5) provided we show that both operators $axv, xva \in L_1(\mathcal{M})$ (the involved operators need not be positive). Clearly, by (3.29) we have $ax \in L_2(\mathcal{M})$, and hence by the Cauchy-Schwarz inequality $axv \in L_1(\mathcal{M})$. For xva note that for each $s > 0$

$$\mu_{3s}(xva) \leq \mu_{2s}(xv)\mu_s(a) \leq \mu_s(x)\mu_s(v)\mu_s(a)$$

(see [17, Lemma 2.5]). A simple calculation shows that $\mu(a) \in E^{\times 2\times 1/2}$ defines a bounded multiplier from $E^{\times}$ into L_2 (see e.g. the argument from [8, Proposition 3.5]), hence we see that $\mu(v), \mu(x)\mu(a) \in L_2$. Again by the Cauchy-Schwarz inequality we get $\mu(xva) \in L_1$, and hence $xva \in L_1(\mathcal{M})$. Now we use (3.31) and (3.29) to prove (3.30):

$$
\begin{aligned}
\sup_{\|u\|_2^2+\|v\|_2^2\leq 1} \|M^a(u,v)\|_{E(\mathcal{M})} &= \sup_{\|u\|_2^2+\|v\|_2^2\leq 1} \sup_{\|x\|_{E^\times(\mathcal{M})}\leq 1} |\tau(x(au+va))| \\
&= \sup_{\|x\|_{E^\times(\mathcal{M})}\leq 1} \sup_{\|u\|_2^2+\|v\|_2^2\leq 1} |\tau(xau)+\tau(xva)| \\
&= \sup_{\|x\|_{E^\times(\mathcal{M})}\leq 1} \sup_{\|u\|_2^2+\|v\|_2^2\leq 1} |\tau(axv)+\tau(xau)| \\
&= \sup_{\|x\|_{E^\times(\mathcal{M})}\leq 1} \|M_a(x)\|_{L_2(\mathcal{M})\oplus_2 L_2(\mathcal{M})} \leq \sqrt{2}\,\|d\|_{E^{\times 2\times}}^{1/2}.
\end{aligned}
$$

From (3.28) we deduce that the definition $R(ax,xa) := w'x, x \in \mathcal{M}$ through continuous extension leads to a well defined, bounded and linear operator

$$R : L_2(\mathcal{M}) \oplus_2 L_2(\mathcal{M}) \longrightarrow H',$$

for which the diagram

$$
\begin{array}{ccc}
E^\times(\mathcal{M}) & \xrightarrow{\quad w' \quad} & H' \\
& \searrow_{M_a} & \uparrow_{R} \\
& & L_2(\mathcal{M}) \oplus_2 L_2(\mathcal{M})
\end{array}
$$

commutes. After dualization we get the commutative diagramm

$$
\begin{array}{ccc}
H & \xrightarrow{\quad w \quad} & E(\mathcal{M}) \\
& \searrow_{(U,V)} & \uparrow_{M^a} \\
& & L_2(\mathcal{M}) \oplus_2 L_2(\mathcal{M}),
\end{array}
\tag{3.32}
$$

where now $(U,V) := R'$. The conclusion of the theorem then follows from (3.26) and (3.32) after defining $u_k := Uv(e_k)$ and $v_k := Vv(e_k)$.

In the second part of the proof we assume that $E = L_1$, hence we have $E(\mathcal{M},\tau) = L_1(\mathcal{M},\tau) = \mathcal{M}_*$. Consider again the dual $w' : \mathcal{M} \longrightarrow H'$ of the operator w which was obtained by factorization in (3.26), and choose again according to (3.24) a state φ on $\mathcal{M}$ such that for all $x \in \mathcal{M}$

$$\|w'x\| \le C\|w\|\varphi\left(\frac{x^*x+xx^*}{2}\right)^{\frac{1}{2}}.\tag{3.33}$$

Equip $\mathcal{M}$ with a semi scalar product

$$(x|y)_\varphi := \varphi\left(\frac{y^*x+xy^*}{2}\right)^{\frac{1}{2}}.$$

Then after passing to the quotient of $\mathcal{M}$ by the kernel $\{x \in \mathcal{M} \mid (x|x)_\varphi = 0\}$ and after completing, we obtain the Hilbert space $L_2(\mathcal{M},\varphi)$. By construction the canonical mapping

$$q_\varphi : \mathcal{M} \longrightarrow L_2(\mathcal{M},\varphi)$$

has norm $\|q_\varphi\| \le 1$, and after defining $Rq_\varphi x := w'x$, $x \in \mathcal{M}$ inequality (3.33) leads to a factorization

$$\tag{3.34}$$

Let $P_n : \mathcal{M}' \longrightarrow \mathcal{M}_*$ be the projection onto the normal part; more precisely, there is a central projection $z \in \mathcal{M}''$ such that for all $x \in \mathcal{M}$ and $\psi \in \mathcal{M}'$

$$\langle P_n\psi, x\rangle = \psi(zx)\tag{3.35}$$

(see [89, p.126]). Put $\varphi_n := P_n\varphi \in \mathcal{M}_* = L_1(\mathcal{M},\varphi)$, and note that the identity map on $\mathcal{M}$ extends to a contraction

$$i_n : L_2(\mathcal{M},\varphi) \longrightarrow L_2(\mathcal{M},\varphi_n);\tag{3.36}$$

indeed, if $\varphi_s = \varphi - \varphi_n \in \mathcal{M}'$ denotes the singular part of φ, then for each $x \in \mathcal{M}$

$$\langle x^*x, \varphi_s\rangle = \langle x^*x, \varphi - \varphi_n\rangle = \varphi((1-z)x^*x)$$

$$= \varphi(x^*(1-z)x) = \varphi\left(((1-z)^{\frac{1}{2}}x)^*(1-z)^{\frac{1}{2}}x\right) \ge 0$$

(for the third equality use the fact that z is central), hence

$$(x|x)_{\varphi_n} \le (x|x)_{\varphi_n} + (x|x)_{\varphi_s} = (x|x)_\varphi.$$

Moreover, the canonical mapping $q_{\varphi_n} : \mathcal{M} \longrightarrow L_2(\mathcal{M},\varphi_n)$ is weak*-weak continuous, so that its dual q'_{φ_n} has values in $\mathcal{M}_*$. Together with the Riesz map

$$R_\varphi : L_2(\mathcal{M},\varphi) \longrightarrow L_2(\mathcal{M},\varphi)'$$

and the contraction

$$S_{\varphi_n} : L_2(\mathcal{M}, \varphi_n) \longrightarrow L_2(\mathcal{M}, \tau) \oplus_2 L_2(\mathcal{M}, \tau), \quad S_{\varphi_n}(x) := \left(\varphi_n^{\frac{1}{2}} x^*, x^* \varphi_n^{\frac{1}{2}} \right)$$

we obtain from (3.34) and (3.36) the following commutative diagram

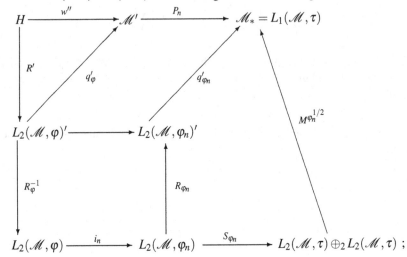

in order to show that $P_n q'_\varphi = q'_{\varphi_n} \left(R_{\varphi_n} i_n R_\varphi^{-1} \right)$ one again has to use the fact that z is central. Since $w = P_n w''$, this is a commutative diagram analog to (3.32) so that the proof finishes exactly as in the first case.   □

As anounced, we finally prove the row+column maximal theorem.

Proof (of Theorem 19). Assume that $E(\mathcal{M}, \tau)$ is a symmetric space of operators (E 2-concave and such that either $E^\times$ is ordercontinuous or $E = L_1$), and let $\sum_k x_k$ be an unconditionally convergent series in $E(\mathcal{M}, \tau)$. Choose a splitting $x_n = a u_n + v_n a$ according to Lemma 19, define $w_n^r := a u_n$ and $w_n^c := v_n a$, and note that we obviously have

$$\sum_k x_k = \sum_k w_k^r + \sum_k w_k^c.$$

Then by Theorem 17 for every $(2,2)$-maximizing matrix $A = (a_{jk})$ we get that

$$\left(\sum_{k=0}^\infty a_{jk} u_k \right)_j \in L_2(\mathcal{M})[\ell_\infty^r] \text{ and } \left(\sum_{k=0}^\infty a_{jk} v_k \right)_j \in L_2(\mathcal{M})[\ell_\infty^c].$$

According to the definition of $L_2[\ell_\infty^c]$ and $L_2[\ell_\infty^r]$ choose now uniform factorizations

$$\sum_{k=0}^\infty a_{jk} u_k = x^r z_j^r \text{ and } \sum_{k=0}^\infty a_{jk} v_k = z_j^c x^c,$$

where for $\alpha = r$ or c both operators x^α are in $L_2(\mathcal{M})$, and both sequences (z_k^α) are uniformly bounded in $\mathcal{M}$. From (3.25) we know that the sequences (w_n^r) and (w_n^c) are unconditionally summable in $E(\mathcal{M}, \tau)$, and moreover by the definition of $E(\mathcal{M})[\ell_\infty^\alpha]$ and again (3.25) we have

$$\left(\sum_{k=0}^\infty a_{jk} w_k^r \right)_j = \left(a x^r z_j^r \right)_j \in E(\mathcal{M})[\ell_\infty^r]$$

$$\left(\sum_{k=0}^\infty a_{jk} w_k^c \right)_j = \left(z_j^c x^c a \right)_j \in E(\mathcal{M})[\ell_\infty^c].$$

This finishes the proof of Theorem 19.   □

3.1.5 Almost Uniform Convergence

Segal [84] was the first who in 1953 introduced and studied "pointwise convergence of operators" in semifinite von Neumann algebras in the spirit of Egoroff's classical characterization of almost everywhere convergence of measurable functions (see (1.18)). Since then it has been used systematically in noncommutative integration theory (see also the beginning of Sect. 3.2.7 for further references).

The first two of the following three definitions on almost uniform convergence for sequences of operators in $L_0(\mathcal{M}, \tau)$ are well-known, the third one seems to be new.

- A sequence (y_n) in $L_0(\mathcal{M}, \tau)$ is said to converge τ-almost uniformly to $y \in L_0(\mathcal{M}, \tau)$ if for each $\varepsilon > 0$ there is $p \in \mathcal{M}_{\mathrm{proj}}$ such that $\tau(1 - p) < \varepsilon$, $(y_n - y)p \in \mathcal{M}$ for all n and $\|(y_n - y)p\|_\infty \to 0$.
- it is said to converge bilaterally τ-almost uniformly whenever in the preceding definition we have $p(y_n - y)p \in \mathcal{M}$ and $\|p(y_n - y)p\|_\infty \to 0$ instead of $(y_n - y)p \in \mathcal{M}$ and $\|(y_n - y)p\|_\infty \to 0$.
- and finally we say that (y_n) converges row+column τ-almost uniformly provided the sequence (y_n) decomposes into a sum $(a_n) + (b_n)$ of two sequences where (a_n) and (b_n^*) both are τ-almost uniformly convergent.

We will later see in Sect. 3.1.8 that here in general only the trivial implications hold: τ-almost uniform convergence implies row+column τ-almost uniform convergence which implies bilateral τ-almost uniform convergence.

Lemma 20. Let $a \in E(\mathcal{M}, \tau)$ with $\|a\|_{E(\mathcal{M})} < 1$. Then for each $0 < \varepsilon < 1$ there is a projection p in $\mathcal{M}$ such that

$$\tau(1 - p) < \varepsilon \text{ and } \|ap\|_\infty < \frac{2}{\varepsilon}.$$

Proof. Recall from (3.2) that there is a contraction

$$E(\mathcal{M}) \hookrightarrow L_1(\mathcal{M}) + L_\infty(\mathcal{M}).$$

Hence by the definition of the norm in the latter space there are $a_1 \in L_1(\mathcal{M})$ and $a_2 \in \mathcal{M}$ such that $a = a_1 + a_2$ and $\|a_1\|_1 + \|a_2\|_\infty < 1$. Clearly

$$\sup_{t>0} t\, \mu_t(a_1) \leq \int_0^\infty \mu_t(a_1)\,dt = \|a_1\|_1 < 1,$$

which by (3.1) yields some projection p with $\tau(1-p) < \varepsilon$ and $\|a_1 p\|_\infty < \frac{1}{\varepsilon}$. But then

$$\|ap\|_\infty \leq \|a_1 p\|_\infty + \|a_2 p\|_\infty < \frac{1}{\varepsilon} + \|a_2\|_\infty < \frac{2}{\varepsilon}.$$

$\square$

The next result is a simple consequence.

Lemma 21. *Every convergent sequence (a_n) in $E(\mathcal{M}, \tau)$ has a τ-almost uniformly convergent subsequence (and hence also a bilaterally τ-almost uniformly convergent subsequence).*

Proof. Clearly, it suffices to check that for every sequence (a_n) in $E(\mathcal{M})$ with $\|a_n\|_E < 2^{-2n}/2$ the series $\sum_n a_n$ converges τ-almost uniformly. For a given $\varepsilon > 0$, choose by Lemma 20 projections p_n such that

$$\tau(1-p_n) < \frac{\varepsilon}{2} 2^{-n} \text{ and } \|a_n p_n\|_\infty < 2^{-2n} \frac{4}{\varepsilon} 2^n = \frac{4}{\varepsilon} 2^{-n}.$$

Then for the projection $p := \inf_n p_n$ we get

$$\tau(1-p) = \tau(\sup_n (1-p_n)) \leq \sum_n \tau(1-p_n) \leq \varepsilon$$

and

$$\|\sum_{n=m}^\infty a_n p\|_\infty \leq \sum_{n=m}^\infty \|a_n p\|_\infty \leq \frac{4}{\varepsilon} \sum_{k=m}^\infty 2^{-k},$$

which tends to 0 whenever m tends to ∞.

$\square$

In our noncommutative setting the next proposition will serve as a perfect substitute of Lemma 2.

Proposition 6. *Assume that the sequence (x_n) from $E(\mathcal{M}, \tau)$ satisfies*

$$(x_n - x_m)_{nm} \in E(\mathcal{M})[c_0(\mathbb{N}_0^2)]. \tag{3.37}$$

Then (x_n) converges to some x in $E(\mathcal{M})$, and

$$x = \lim_n x_n \text{ bilaterally } \tau\text{-almost uniformly;}$$

if in (3.37) we may replace c_0 by c_0^c, then bilateral τ-almost uniform convergence can be replaced by τ-almost uniform convergence, and if we have c_0^{r+c} instead of c_0, then it is possible to replace bilateral τ-almost uniform convergence by row+column τ-almost uniform convergence.

Proof. We show the symmetric case; the proof of the column case is similar, and finally the proof of the row+column case is a simple consequence. Fix some $0 < \varepsilon < 1$. By assumption there is a factorization

$$x_n - x_m = a_1 u_{nm} a_2$$

with $a_1, a_2 \in E^{1/2}(\mathscr{M})$ and $\lim_{nm} \|u_{nm}\|_\infty = 0$. Without loss of generality we may assume that $\|a_1\|_{E^{1/2}(\mathscr{M})}, \|a_2\|_{E^{1/2}(\mathscr{M})} < 1$, and hence by Lemma 20 there are two projections $p_k \in \mathscr{M}_{\mathrm{proj}}$ with $\tau(1 - p_k) < \varepsilon$ and

$$\max\{\|p_1 a_1\|_\infty, \|a_2 p_2\|_\infty\} < 2/\varepsilon.$$

In particular, we have for $p = \inf\{p_1, p_2\}$ that

$$\tau(1 - p) \leq \tau(1 - p_1) + \tau(1 - p_2) < 2\varepsilon$$

and

$$\|p(x_n - x_m)p\|_\infty \leq \|p_1(x_n - x_m)p_2\|_\infty \leq \frac{4}{\varepsilon^2}\|u_{nm}\|_\infty \to 0 \ \text{ for } n, m \to \infty.$$

On the other hand (x_n) is obviously a Cauchy sequence in the Banach space $E(\mathscr{M})$ which hence converges to some $x \in E(\mathscr{M})$. By Lemma 21 there is a subsequence (x_{n_k}) which converges τ-almost uniformly to x. In particular there is a projection q for which $\tau(1 - q) < \varepsilon$ and $\lim_k \|(x_{n_k} - x)q\|_\infty = 0$. This finally shows that for $s = \inf\{p, q\}$ we have

$$\tau(1 - s) < 3\varepsilon$$

and

$$\lim_n \|s(x_n - x)s\|_\infty = 0,$$

the conclusion.   □

3.1.6   Coefficient Tests in Symmetric Operator Spaces

The following theorems combine several of the preceding results – they form the core of our study of classical summation processes of unconditional convergent series in symmetric spaces $E(\mathscr{M}, \tau)$ build with respect to a semifinite von Neumann algebra and a normal, faithful and semifinite trace.

We will frequently use the following noncommutative analog of Lemma 7 which is the crucial link between our maximal inequalities and τ-almost uniform convergence.

Lemma 22. *Let $A = (a_{jk})$ be an infinite matrix which converges in each column and satisfies $\|A\|_\infty < \infty$. Let $1 \leq q \leq \infty$, and assume that*

$$\left(\sum_{k=0}^{\infty} a_{jk} \alpha_k x_k \right)_j \in E(\mathcal{M}, \tau)[\ell_\infty] \tag{3.38}$$

for every sequence $(\alpha_k) \in \ell_q$ and every weakly q'-summable sequence (x_k) in $E(\mathcal{M}, \tau)$ (in the case $q = \infty$ we only consider unconditionally summable sequences). Then for every such (α_k) and (x_k) the sequence $\left(\sum_{k=0}^{\infty} a_{jk} \alpha_k x_k \right)_j$ converges to some $s \in E(\mathcal{M}, \tau)$, and

(1) $s = \lim_j \sum_{k=0}^{\infty} a_{jk} \alpha_k x_k$ bilaterally τ-almost uniformly
(2) If the assumption in (3.38) holds for ℓ_∞^c instead of ℓ_∞, then the convergence in (1) is even τ-almost uniform.
(3) If the assumption in (3.38) holds for ℓ_∞^{r+c} instead of ℓ_∞, then the convergence in (1) is row+column τ-almost uniform.

Proof. For the proof of (1) we need that for all sequences (α_k) in ℓ_q and (x_k) in $E(\mathcal{M}, \tau)$

$$\left(\sum_{k=0}^{\infty} a_{ik} \alpha_k x_k - \sum_{k=0}^{\infty} a_{jk} \alpha_k x_k \right)_{(i,j)} \in E(\mathcal{M}, \tau)[c_0(\mathbb{N}_0^2)]; \tag{3.39}$$

then the conclusion is an immediate consequence of Proposition 6 from the preceding section. But the statement in (3.39) follows from a simple analysis of the proof of (2.26). Since $E(\mathcal{M}, \tau)[\ell_\infty^c]$ and $E(\mathcal{M}, \tau)[\ell_\infty^{r+c}]$ are quasi Banach spaces, the proofs of (2) and (3) work exactly the same way (recall in particular that the closed graph theorem needed in the proof of (2.26) is valid in quasi Banach spaces). $\square$

We start with a noncommutative variant of Theorem 13 which collects many of the results proved so far (for the definition of Weyl sequences see again (1.6)).

Theorem 20. *Assume that S is a summation method and ω a Weyl sequence with the additional property that for each orthonormal series $\sum_k \alpha_k x_k$ in $L_2(\mu)$ we have*

$$\sup_j \left| \sum_{k=0}^{\infty} s_{jk} \sum_{\ell=0}^{k} \frac{\alpha_\ell}{\omega_\ell} x_\ell \right| \in L_2(\mu).$$

Then for each unconditionally convergent series $\sum_k x_k$ in $E(\mathcal{M},\tau)$ the following statements hold:

(1) $\left(\sum_{k=0}^{\infty} s_{jk}\sum_{\ell=0}^{k}\dfrac{x_\ell}{\omega_\ell}\right)_j \in E(\mathcal{M})[\ell_\infty]$

(2) $\sum_{k=0}^{\infty}\dfrac{x_k}{\omega_k} = \lim_j \sum_{k=0}^{\infty} s_{jk}\sum_{\ell=0}^{k}\dfrac{x_\ell}{\omega_\ell}$ *bilaterally τ-almost uniformly.*

Moreover:

(3) If E is 2-convex, then the sequence in (1) belongs to $E(\mathcal{M})[\ell_\infty^c]$, and the convergence in (2) is even τ-almost uniform.

(4) If E is 2-concave and such that either $E^\times$ is ordercontinuous or $E = L_1$, then the series $\sum_k x_k$ splits into a sum

$$\sum_k x_k = \sum_k u_k + \sum_k v_k$$

of two unconditionally convergent series in $E(\mathcal{M},\tau)$ for which

$$\left(\sum_{k=0}^{\infty} s_{jk}\sum_{\ell=0}^{k}\dfrac{\alpha_\ell}{\omega_\ell}u_\ell\right)_j \in E(\mathcal{M})[\ell_\infty^r] \ , \ \left(\sum_{k=0}^{\infty} s_{jk}\sum_{\ell=0}^{k}\dfrac{\alpha_\ell}{\omega_\ell}v_\ell\right)_j \in E(\mathcal{M})[\ell_\infty^c].$$

In this case, the convergence in (2) is row+column τ-almost uniform.

Proof. Recall that $A = S\Sigma D_{1/\omega}$ is $(2,2)$-maximizing (Theorem 1) and converges in each column (see (2.67)). Hence, the maximal theorems follow from Theorem 17 and Theorem 19 (see also (3.23)), and the results on convergence are then all consequences of Lemma 22. □

As in the preceding chapter we apply our result to Riesz, Cesàro, and Abel summation – for the proof compare with its analogs from Corollary 6.

Corollary 10. *Let $\sum_k x_k$ be an unconditionally convergent series in $E(\mathcal{M},\tau)$. Then*

(1) $\left(\sum_{k=0}^{j}\dfrac{x_k}{\log k}\right)_j \in E(\mathcal{M})[\ell_\infty]$

(2) $\left(\sum_{k=0}^{j}\dfrac{\lambda_{k+1}-\lambda_k}{\lambda_{j+1}}\sum_{\ell\le k}\dfrac{x_\ell}{\log\log\lambda_\ell}\right)_j \in E(\mathcal{M})[\ell_\infty]$ *for every strictly increasing, unbounded and positive sequence (λ_k) of scalars*

(3) $\left(\sum_{k=0}^{j}\dfrac{A_{j-k}^{r-1}}{A_j^r}\sum_{\ell\le k}\dfrac{x_\ell}{\log\log\ell}\right)_j \in E(\mathcal{M})[\ell_\infty]$ *for every $r > 0$*

(4) $\left(\sum_{k=0}^{\infty}\rho_j^k\dfrac{x_k}{\log\log k}\right)_j \in E(\mathcal{M})[\ell_\infty]$ *for every positive strictly increasing sequence (ρ_j) converging to 1*

(5) If E is 2-convex, then ℓ_∞ in $(1) - (3)$ may be replaced by ℓ_∞^c, and if E is 2-concave and such that either $E^\times$ is ordercontinuous or $E = L_1$, then in all four statements there is a decomposition

$$\sum_k s_{jk} \sum_{\ell \leq k} \frac{x_\ell}{\omega_\ell} = \sum_k s_{jk} \sum_{\ell \leq k} \frac{u_\ell}{\omega_\ell} + \sum_k s_{jk} \sum_{\ell \leq k} \frac{v_\ell}{\omega_\ell},$$

where the first summand belongs to $E(\mathcal{M})[\ell_\infty^r]$ and the second one to $E(\mathcal{M})[\ell_\infty^c]$ (in other terms, ℓ_∞ in $(1) - (3)$ may be replaced by ℓ_∞^{r+c}).

Moreover, in $(1) - (4)$

$$\sum_{k=0}^\infty \frac{x_k}{\omega_k} = \lim_j \sum_{k=0}^\infty s_{jk} \sum_{\ell \leq k} \frac{x_\ell}{\omega_\ell} \quad \textit{bilaterally } \tau\textit{-almost uniformly};$$

if E is 2-convex, then this convergence is τ-almost uniform, and if E is 2-concave and such that either $E^\times$ is ordercontinuous or $E = L_1$, then each sequence in $(1) - (4)$ converges row+column τ-almost uniformly.

Within the setting of noncommutative L_p-spaces it is possible to extend Theorem 20. We prove the following noncommutative coefficient test which in contrast to Theorem 20 does not involve any log-terms (an analog of Theorem 14).

Theorem 21. *Let S be a summation method and $1 \leq q < p < \infty$. Then for each $\alpha \in \ell_q$ and each weakly q'-summable sequence (x_k) in $L_p(\mathcal{M},\tau)$ we have that*

(1) $\left(\sum_{k=0}^\infty s_{jk} \sum_{\ell \leq k} \alpha_\ell x_\ell \right)_j \in L_p(\mathcal{M})[\ell_\infty]$

(2) $\sum_{k=0}^\infty \alpha_k x_k = \lim_j \sum_{k=0}^\infty s_{jk} \sum_{\ell \leq k} \alpha_\ell x_\ell$ *bilaterally τ-almost uniformly;*

if $2 \leq p$, then here ℓ_∞ may be replaced by ℓ_∞^c, and bilateral τ-almost uniform by τ-almost uniform convergence.

Proof. From Theorem 2 we know that the matrix $A = S\Sigma$ is (p,q)-maximizing, and clearly it is convergent in each column (see also (2.67)). Then it follows from Theorem 18 that for every $\alpha \in \ell_q$ and every weakly q'-summable sequence (x_k) in $L_p(\mathcal{M},\tau)$ we have

$$\left(\sum_{k=0}^\infty s_{jk} \sum_{\ell \leq k} \alpha_\ell x_\ell \right)_j = \left(\sum_{k=0}^\infty a_{jk} x_k \right)_j \in L_p(\mathcal{M})[\ell_\infty].$$

Again we deduce from Lemma 22 that for every such (α_k) and (x_k) the sequence $\left(\sum_{k=0}^\infty a_{jk} x_k \right)_j$ converges to some $s \in L_p(\mathcal{M},\tau)$, and

$$s = \lim_j \sum_{k=0}^\infty a_{jk} x_k \text{ bilaterally } \tau\text{-almost uniformly};$$

the column case follows the same way. □

We do not know whether the preceding theorem can be improved in the case $p \leq 2$ as in Theorem 20, (4). If we apply it to ordinary summation or Cesàro, Riesz, and Abel summation, then we obtain an analog of Corollary 7.

Corollary 11. *Assume that $1 \leq q < p < \infty$. Then for each $\alpha \in \ell_q$ and each weakly q'-summable sequence (x_k) in $L_p(\mathcal{M}, \tau)$ we have*

(1) $\left(\Sigma_{k=0}^{j} \alpha_k x_k \right)_j \in L_p(\mathcal{M})[\ell_\infty]$

(2) $\left(\Sigma_{k=0}^{j} \dfrac{\lambda_{k+1} - \lambda_k}{\lambda_{j+1}} \Sigma_{\ell \leq k} \alpha_k x_k \right) \in L_p(\mathcal{M})[\ell_\infty]$ *for every strictly increasing, unbounded and positive sequence (λ_k) of scalars*

(3) $\left(\Sigma_{k=0}^{j} \dfrac{A_{j-k}^{r-1}}{A_j^r} \Sigma_{\ell \leq k} \alpha_k x_k \right)_j \in L_p(\mathcal{M})[\ell_\infty]$ *for every $r > 0$*

(4) $\left(\Sigma_{k=0}^{\infty} \rho_j^k \alpha_k x_k \right)_j \in L_p(\mathcal{M})[\ell_\infty]$ *for every positive strictly increasing sequence (ρ_j) converging to 1*

(5) *If $2 \leq p < \infty$, then in $(1) - (4)$ ℓ_∞ may be replaced by ℓ_∞^c.*

Moreover, in $(1) - (4)$

$$\sum_{k=0}^{\infty} \alpha_k x_k = \lim_j \sum_{k=0}^{\infty} s_{jk} \sum_{\ell \leq k} \alpha_\ell x_\ell \quad \text{bilaterally } \tau\text{-almost uniformly;}$$

and if $2 \leq p < \infty$, then this convergence is even τ-almost uniform.

We finish with a noncommutative extension of Corollary 8 – a sort of converse of Corollary 10, (2).

Corollary 12. *Let $\Sigma_k x_k$ be an unconditionally convergent series in $E(\mathcal{M}, \tau)$ where E has finite cotype, and let x be its sum. Then there is a Riesz matrix $R^\lambda = (r_{jk}^\lambda)$ such that*

$$\left(\sum_{k=0}^{\infty} r_{jk}^\lambda \sum_{\ell \leq k} x_\ell \right)_j \in E(\mathcal{M})[\ell_\infty]$$

and

$$\lim_j \sum_{k=0}^{\infty} r_{jk}^\lambda \sum_{\ell \leq k} x_\ell = x \quad \text{bilaterally } \tau\text{-almost uniformly.} \tag{3.40}$$

Moreover, ℓ_∞ may be replaced by ℓ_∞^c and the convergence is τ-almost uniform provided that E is 2-convex. If E is 2-concave and such that $E^\times$ is ordercontinuous or $E = L_1$, then R^λ may be chosen in such a way that the equality in (3.40) holds row+column almost uniformly; corresponding maximal inequalities hold.

The proof is absolutely the same as that of Corollary 8; recall that $E(\mathcal{M}, \tau)$ has finite cotype whenever E has (see [18, Theorem 4] and [95]), and use as before the Theorems 8 and 20.

3.1.7 Laws of Large Numbers in Symmetric Operator Spaces

Classical coefficient tests for orthonormal series via Kronecker's lemma (see Lemma 11) induce laws of large numbers. We now continue Sect. 2.3.2 where we showed how to transfer such strong laws of large numbers for sequences of uncorrelated random variables to unconditionally summable sequences in vector-valued Banach function spaces. We prove that these laws even can be transferred to the setting of symmetric spaces of operators. Our first theorem is a noncommutative analog of Theorem 15.

Theorem 22. *Let S be a lower triangular summation method. Assume that ω is an increasing sequence of positive scalars such that for each orthogonal sequence (x_k) in $L_2(\mu)$ with $\sum_k \frac{\omega_k^2}{k^2}\|x_k\|_2^2 < \infty$ we have*

$$\sup_j \left| \frac{1}{j+1} \sum_{k=0}^{j} s_{jk} \sum_{\ell=0}^{k} x_\ell \right| \in L_2(\mu).$$

Then for each unconditionally convergent series $\sum_k \frac{\omega_k}{k} x_k$ in $E(\mathcal{M},\tau)$ the following two statements hold:

(1) $\left(\frac{1}{j+1} \sum_{k=0}^{j} s_{jk} \sum_{\ell=0}^{k} x_\ell \right)_j \in E(\mathcal{M})[\ell_\infty]$

(2) $\lim_j \frac{1}{j+1} \sum_{k=0}^{j} s_{jk} \sum_{\ell=0}^{k} x_\ell = 0$ *bilaterally τ-almost uniformly.*

Moreover:

(3) If E is 2-convex, then the sequence in (1) is in $E(\mathcal{M})[\ell_\infty^c]$, and the series in (2) converges even τ-almost uniformly.

(4) If E is 2-concave and such that either $E^\times$ is ordercontinuous or $E = L_1$, then there is a decomposition

$$\frac{1}{j+1} \sum_k s_{jk} \sum_{\ell \le k} x_\ell = \frac{1}{j+1} \sum_k s_{jk} \sum_{\ell \le k} u_\ell + \frac{1}{j+1} \sum_k s_{jk} \sum_{\ell \le k} v_\ell,$$

where the first summand belongs to $E(\mathcal{M})[\ell_\infty^r]$ and the second one to $E(\mathcal{M})[\ell_\infty^c]$. Moreover, the convergence in (2) is even row+column τ-almost uniform.

Proof. Of course the proof is very similar to that of Theorem 15. We first conclude from the assumption and Theorem 1 that

$$b_{jk} := \begin{cases} \dfrac{k}{(j+1)\omega_k} \sum_{\ell=k}^{j} s_{j\ell} & k \le j \\ 0 & k > j \end{cases}$$

is $(2,2)$-maximizing, and moreover we have for every choice of elements $y_0, \ldots, y_j$ in any vector space that

$$\sum_{k=0}^{j} b_{jk} y_k = \sum_{k=0}^{j} \frac{1}{j+1} s_{jk} \sum_{\ell=0}^{k} \frac{\ell}{\omega_\ell} y_\ell .$$

Since $\sum_k \frac{\omega_k}{k} x_k$ is unconditionally convergent in $E(\mathcal{M}, \tau)$, we conclude from Theorem 17 and Theorem 19 all the stated maximal inequalities, and then from Lemma 22 that $\left(\frac{1}{j+1} \sum_{k=0}^{\infty} s_{jk} \sum_{\ell=0}^{k} x_\ell \right)_j$ converges to its $E(\mathcal{M})$-limit s bilaterally τ-almost uniformly, τ-almost uniformly, and row+column τ-almost uniformly, respectively (note that B converges in each column, compare with (2.67)). It remains to show that $s = 0$. But for the matrix $A = S\Sigma$ we have

$$\lim_j \sum_{k=0}^{j} a_{jk} \frac{x_k}{k} = \lim_j \sum_{k=0}^{j} s_{jk} \sum_{\ell=0}^{k} \frac{x_\ell}{\ell} = \sum_{k=0}^{\infty} \frac{x_k}{k},$$

the limits taken in $E(\mathcal{M})$, and hence by Kronecker's Lemma 11, (2) we see that in $E(\mathcal{M})$

$$0 = \lim_j \frac{1}{j+1} \sum_{k=0}^{j} a_{jk} x_k = \lim_j \frac{1}{j+1} \sum_{k=0}^{j} s_{jk} \sum_{\ell=0}^{k} x_\ell .$$

This completes the proof.   □

If combined with (2.72), then the preceding result implies as an immediate consequence the following noncommutative variant of Corollary 9.

Corollary 13. *For sequences* (x_k) *in* $E(\mathcal{M}, \tau)$ *for which* $\sum_k \frac{\log k}{k} x_k$ *converges unconditionally, we have*

$$\lim_j \frac{1}{j+1} \sum_{k=0}^{j} x_k = 0 \quad \text{bilaterally } \tau-\text{almost uniformly}$$

and

$$\left(\sum_{k=0}^{j} x_k \right)_j \in E(\mathcal{M})[\ell_\infty].$$

Clearly, this result improves provided E is 2-convex or 2-concave. Similar laws of large numbers can be deduced for Riesz, Cesàro, and Abel summation; combine Lemma 14 with Theorem 22 and Theorem 8 (Riesz), Theorem 10 (Cesàro), as well as Theorem 12 (Abel). But the following noncommutative counterpart of Theorem 16 again shows that for Cesàro summation of order $r > 0$ no logarithmical terms are needed.

Theorem 23. *Let* $\sum_k \frac{x_k}{k}$ *be an unconditionally convergent series in* $E(\mathcal{M}, \tau)$ *and* $r > 0$. *Then we have that*

$$\lim_j \frac{1}{j+1} \sum_{k=0}^{j} \frac{A_{j-k}^{r-1}}{A_j^r} \sum_{\ell \leq k} x_\ell = 0 \quad \text{bilaterally } \tau\text{-almost uniformly;}$$

*if E is 2-convex, then this convergence is τ-almost uniform, and if E is 2-concave
and such that either $E^\times$ is ordercontinuous or $E = L_1$, then the convergence is
row+column almost uniform.*
Related maximal inequalities hold:

$$\left(\frac{1}{j+1}\sum_{k=0}^{j}\frac{A_{j-k}^{r-1}}{A_j^r}\sum_{\ell\leq k}x_\ell\right)_j \in E(\mathcal{M})[\ell_\infty];$$

*if E is 2-convex, then ℓ_∞ may be replaced by ℓ_∞^c, and if E is 2-concave and such that
either $E^\times$ is ordercontinuous or $E = L_1$, then the above sequence decomposes into
a sum*

$$\frac{1}{j+1}\sum_k\frac{A_{j-k}^{r-1}}{A_j^r}\sum_{\ell\leq k}u_\ell+\frac{1}{j+1}\sum_k\frac{A_{j-k}^{r-1}}{A_j^r}\sum_{\ell\leq k}v_\ell,$$

*where the first summand is in $E(\mathcal{M})[\ell_\infty^r]$ and the second one in $E(\mathcal{M})[\ell_\infty^c]$.
In particular, these results hold in the case $r = 1$ which means ordinary Cesàro
summation.*

Proof. By Theorem 16 we know that for each orthogonal sequence (x_k) in $L_2(\mu)$
with $\sum_k\frac{\omega_k^2}{k^2}\|x_k\|_2^2 < \infty$ we have

$$\sup_j\left|\frac{1}{j+1}\sum_{k=0}^{j}\frac{A_{j-k}^{r-1}}{A_j^r}\sum_{\ell=0}^{k}x_\ell\right| \in L_2(\mu).$$

hence the conclusion follows from Theorem 22.   □

3.1.8 A Counterexample

In this section we give a counterexample (taken from [7, Sect. 6.5]) which shows
that in the preceding results (in particular in Theorem 20 and Corollary 10) bilateral
almost uniform convergence in general cannot be replaced by almost uniform
convergence. Clearly, such an example cannot be found in the Schatten class S_p
since here trivially each convergent sequence converges in $\mathcal{L}(\ell_2)$, hence tr-almost
uniformly (tr the natural trace on $\mathcal{L}(\ell_2)$).

Denote by M_n the algebra of all scalar $n \times n$ matrices together with its normalized
trace $\tau_n(x) := \frac{1}{n}\sum_k x_{kk}$. Let us recall the definition of the hyperfinite factor R and its
natural trace τ_R constructed over the algebra M_2: Clearly, for $1 \leq p < \infty$ the identity
maps

$$L_p(M_{2^n},\tau_n) \longrightarrow S_p^{2^n},x \rightsquigarrow \frac{1}{2^{n/p}}x$$

are metric bijections where $S_p^{2^n}$ as usual denotes the Schatten p-class of dimension
2^n (on the Hilbert space $\ell_2^{2^n-1}$; recall from the beginning of Sect. 2.1.2 and the
preliminaries that our ℓ_p^m's are $m + 1$-dimensional). The inflation maps

$$M_{2^n} \longrightarrow M_{2^{n+1}}, \; \square \rightsquigarrow \begin{smallmatrix}\square\\\square\end{smallmatrix}$$

are unitial, and induce metric injections $j_{nm} : M_{2^n} \to M_{2^m}$, $n \leq m$. In particular, the mappings

$$j_{nm} : L_2(M_{2^n}, \tau_n) \longrightarrow L_2(M_{2^m}, \tau_m)$$

are metric injections and lead to the directed limit $H := \lim_n L_2(M_{2^n}, \tau_n)$, a Hilbert space. For $y \in M_{2^n}$ and $x \in M_{2^m}$ define yx to be $(j_{nm}y)x \in M_{2^m}$ if $n \leq m$, and $y(j_{mn}x) \in M_{2^n}$ if $m \leq n$. Now let R be the von Neumann algebra given by the weak closure of all (extensions of) left multipliers

$$H \to H, \; x \rightsquigarrow yx, \; y \in \bigcup_n M_{2^n}.$$

By extension there is a faithful, finite and normal trace τ_R on R which satisfies

$$\tau_R(y) = \tau_n(y) \text{ for all } y \in M_{2^n}.$$

Example 3. There is a sequence (x_j) in R which in $L_1(R, \tau_R)$ is unconditionally summable, but such that for each projection $p \in R_{\text{proj}}$ with $\tau_R(p) > 0$ there is $\xi \in pH$ for which the series

$$\sum_j \frac{x_j(\xi)}{\log j}$$

is not convergent in H. In particular, the unconditionally convergent series

$$\sum_j \frac{x_j}{\log j}$$

from $L_1(R, \tau_R)$ is not τ_R-almost uniformly convergent. In contrast to this, note that the series $\sum_{j=0}^{\infty} \frac{x_j}{\log j}$ by Corollary 10 is bilaterally τ_R-almost uniformly convergent, and even splits into a sum $(a_n) + (b_n)$ where both sequences (a_n^*) and (b_n) converge τ_R-almost uniformly.

As a byproduct this example also shows that almost uniform convergence of a sequence of operators in general does not imply that the adjoint sequence converges almost uniformly. Moreover, an analysis of our construction gives slightly more: For each $1 \leq p < 2$ and $1 \leq q \leq 2$ there exists an unconditionally convergent series $\sum_{j=0}^{\infty} \alpha_j x_j$ in $L_p(R, \tau_R)$ which is not τ_R-almost uniformly convergent although (x_j) is weakly q'-summable and $(\alpha_j \log j) \in \ell_q$.

Proof. For $n \in \mathbb{N}$ and $1 \leq i \leq 2^n$ put $c_n := n^{-2} 2^{n/2}$ and

$$x_{n,i} := c_n e_{1i} \in M_{2^n} \subset R \subset L_1(R).$$

Using the following natural ordering these matrices define a sequence $(x_j)_{j=0}^\infty$ in R:

$$0, x_{1,1}, x_{1,2}, x_{2,1}, \ldots, x_{2,2^2}, \ldots, x_{n,1}, \ldots, x_{n,2^n}, \ldots \ldots;$$

denote the index of the ith element, $1 \le i \le 2^n$, in the nth block by $j = k(n,i)$. Note first that (x_j) as a sequence of $L_1(R)$ is unconditionally summable: For each sequence $(\varepsilon_{n,i})_{n,i}$ of scalars with $|\varepsilon_{n,i}| \le 1$ we have

$$\left\| \sum_{n,i} \varepsilon_{n,i} x_{n,i} \right\|_1 \le \sum_{n=1}^\infty c_n \left\| \sum_{i=1}^{2^n} \varepsilon_{n,i} e_{1i} \right\|_{L_1(M_{2^n})}$$

$$= \sum_{n=1}^\infty c_n \frac{1}{2^n} \left\| \sum_{i=1}^{2^n} \varepsilon_{n,i} e_{1i} \right\|_{S_1^{2^n}}$$

$$= \sum_{n=1}^\infty n^{-2} 2^{-n/2} \left\| \sum_{i=1}^{2^n} \varepsilon_{n,i} e_{i-1} \right\|_{\ell_2^{2^n-1}} = \sum_{n=1}^\infty n^{-2} < \infty.$$

Now assume, contrary to what we intend to prove, that there is some projection $p \in R_{\mathrm{proj}}$ with $\tau_R(p) > 0$ such that the sequence

$$\left(\sum_{j=0}^m \frac{x_j(\xi)}{\log j} \right)_m$$

of partial sums for all $\xi \in pH$ is a Cauchy sequence in H. Then for all $\xi \in H$

$$\sup_{k \le l} \left\| \sum_{j=k}^l \frac{x_j(p\xi)}{\log j} \right\|_H < \infty.$$

Hence, since all x_j are bounded operators, by the uniform boundedness principle

$$\sup_{k \le l} \sup_{\|\xi\| \le 1} \left\| \sum_{j=k}^l \frac{x_j(p\xi)}{\log j} \right\|_H < \infty,$$

which by taking adjoints, gives that

$$c := \sup_{k \le l} \left\| p \sum_{j=k}^l \frac{x_j^*}{\log j} \right\|_\infty = \sup_{k \le l} \left\| \sum_{j=k}^l \frac{x_j}{\log j} p \right\|_\infty < \infty.$$

In particular, for all n and $1 \le r \le 2^n$

$$\left\| p \sum_{i=1}^r \frac{x_{n,i}^*}{\log k(n,i)} \right\|_\infty \le c,$$

so that

$$\left\| p \sum_{i=1}^{r} \frac{1}{\log k(n,i)} e_{i1} \right\|_{\infty} \leq cc_n^{-1}.$$

If now π_{1r} denotes the matrix in M_{2^n} which interchanges the first and the rth coordinate of $\xi \in \mathbb{C}^{2^n}$, then we obtain

$$\left\| p \sum_{i=1}^{r} \frac{1}{\log k(n,i)} e_{ir} \right\|_{\infty} = \left\| p \sum_{i=1}^{r} \frac{1}{\log k(n,i)} e_{i1} \pi_{1r} \right\|_{\infty}$$

$$\leq \left\| p \sum_{i=1}^{r} \frac{1}{\log k(n,i)} e_{i1} \right\|_{\infty} \leq cc_n^{-1}. \tag{3.41}$$

We will show that (3.41) leads to a contradiction: Consider again the sum operator

$$s_{2^n}(\xi) := \left(\sum_{i=1}^{r} \frac{1}{\log k(n,i)} \xi_i \right)_{r \leq 2^n}, \quad \xi \in \mathbb{C}^{2^n}$$

and check as in (3.20) and (3.21) that for all $1 \leq r \leq 2^n$

$$s_{2^n}^* e_{rr} = \sum_{i=1}^{r} \frac{1}{\log k(n,i)} e_{ir} \tag{3.42}$$

and

$$\| (s_{2^n}^*)^{-1} : \ell_2^{2^n-1} \to \ell_2^{2^n-1} \|_{\infty} \leq 2 \sup_{1 \leq i \leq 2^n} \log k(n,i) \leq 8n. \tag{3.43}$$

Now take a projection $q \in R_{\text{proj}}$ onto the range of $s_{2^n} pH$. Obviously,

$$q s_{2^n} p = s_{2^n} p \quad \text{and} \quad qH = s_{2^n} pH.$$

Since then

$$pH = s_{2^n}^{-1}(s_{2^n} pH) = s_{2^n}^{-1} qH,$$

we also have

$$p s_{2^n}^{-1} q = s_{2^n}^{-1} q,$$

and taking adjoints we see that

$$p s_{2^n}^* q = p s_{2^n}^*$$

$$q (s_{2^n}^*)^{-1} p = q (s_{2^n}^*)^{-1}. \tag{3.44}$$

Then by (3.44), (3.43), (3.42), and (3.41)

$$
\begin{aligned}
\sup_{1 \leq r \leq 2^n} \|qe_{rr}\|_\infty &= \sup_{1 \leq r \leq 2^n} \|q(s_{2^n}^*)^{-1} s_{2^n}^* e_{rr}\|_\infty \\
&= \sup_{1 \leq r \leq 2^n} \|q(s_{2^n}^*)^{-1} p s_{2^n}^* e_{rr}\|_\infty \\
&= \|q(s_{2^n}^*)^{-1}\|_\infty \sup_{1 \leq r \leq 2^n} \|p s_{2^n}^* e_{rr}\|_\infty \\
&\leq 8n \sup_{1 \leq r \leq 2^n} \left\| p \sum_{i=1}^{r} \frac{1}{\log k(n,i)} e_{ir} \right\|_\infty \leq 8ncc_n^{-1} = 8cn^3 2^{-n/2}.
\end{aligned}
$$

Let E_n be the conditional expectation of R onto M_{2^n}. Then by the module property of E_n we have

$$
\sup_{1 \leq r \leq 2^n} \|E_n(q)e_{rr}\|_\infty = \sup_{1 \leq r \leq 2^n} \|E_n(qe_{rr})\|_\infty ,
$$

hence

$$
\begin{aligned}
\|E_n(q) : \ell_2^{2^n-1} \to \ell_2^{2^n-1}\|_{S_2^{2^n}} &\leq 2^{n/2} \|E_n(q) : \ell_1^{2^n-1} \to \ell_2^{2^n-1}\| \\
&= 2^{n/2} \sup_{1 \leq r \leq 2^n} \|E_n(q)e_{r-1}\| \\
&\leq 2^{n/2} \sup_{1 \leq r \leq 2^n} \|E_n(q)e_{rr}e_{r-1}\| \\
&\leq 2^{n/2} 8cn^3 2^{-n/2} = 8cn^3 .
\end{aligned}
$$

Finally, we calculate a lower bound of $\|E_n(q)\|_{S_2^{2^n}}$ in order to obtain a contradiction. Note first that by (3.44)

$$
\begin{aligned}
\tau_R(q) &= \tau_R(s_{2^n}^* q(s_{2^n}^*)^{-1}) = \tau_R(s_{2^n}^* q(s_{2^n}^*)^{-1} p) \\
&= \tau_R(p s_{2^n}^* q(s_{2^n}^*)^{-1}) = \tau_R(p s_{2^n}^* (s_{2^n}^*)^{-1}) = \tau_R(p),
\end{aligned}
$$

hence, since E_n is trace preserving, also

$$
\|E_n(q)\|_{S_1^{2^n}} = 2^n \tau_R(E_n(q)) = 2^n \tau_R(q) = 2^n \tau_R(p).
$$

But then for all n

$$
2^n \tau_R(p) = \|E_n(q)\|_{S_1^{2^n}} \leq 2^{n/2} \|E_n(q)\|_{S_2^{2^n}} \leq 2^{n/2} 8cn^3 ,
$$

which, since $\tau_R(p) > 0$, clearly is a contradiction.   □

3.2   The Nontracial Case

So far we focused our attention on the semifinite case only. Keeping the same strategy, in this section the previous theory will be extended to arbitrary noncommutative probability spaces $(\mathcal{M}, \varphi)$, i.e. $\mathcal{M}$ a finite von Neumann algebra equipped with a distinguished normal faithful state φ. As in the tracial case our main aim is to provide a general setting which allows to deduce from classical commutative coefficient tests their noncommutative analogs in noncommutative Haagerup L_p-spaces. We continue, extend and supplement many results obtained in the eighties and nineties by the school of Łódź, mainly by Hensz and Jajte, and which were later collected in the two lecture notes [37] and [38].

In a first step we again provide proper maximal theorems. The three main extension theorems for maximizing matrices in the tracial case were Theorem 17, 18, and 19. In Haagerup L_p-spaces we get Theorems 24 and 25 as substitutes of these results. As in the tracial case (see Sect. 3.1.2) our main tool comes from ℓ_∞-valued variants of Haagerup L_p-spaces which allow to formulate proper maximal theorems (inequalities).

In a second step we deduce from these maximal theorems almost everywhere convergence theorems – our study here is based on the definition of almost sure convergence for operators in the Hilbert space $L_2(\mathcal{M}, \varphi)$ (GNS-construction) which was first defined by Hensz and Jajte in [33] (and is again motivated by Egoroff's theorem). In the setting of type III von Neumann algebras it is technically much more complicated to deduce from our maximal inequalities results on almost sure convergence. This is mainly due to the fact that for two projections p and q the inequality $\varphi(p \vee q) \leq \varphi(p) + \varphi(q)$ is no longer true. Many typically set theoretic arguments which in the semifinite case still have a clear meaning, cannot be used in the more general nontracial case. Mainly because of Goldstein's Chebychev type inequality from [19] (see Lemma 24) we here are able to overcome these difficulties. The abstract coefficient tests from Theorems 26 and 27 in Haagerup L_p-spaces lead to almost perfect analogs of the ones we had obtained in the tracial case; as before we apply our results to abstract as well as concrete summation methods.

In Sect. 3.2.7 we discuss our program for sequences in von Neumann algebras $\mathcal{M}$ itself. We adopt and extend the very natural definition of almost uniform convergence in von Neumann algebras given by Lance in [51] (and which was already repeated in the introduction), and finish with corresponding maximal theorems.

3.2.1   Haagerup L_p-Spaces

The noncommutative L_p-spaces used below will be those constructed by Haagerup in [24], and as usual our general reference will be [90] (see the end of this section for a short list of further references). Although we will use Haagerup L_p's in a quite axiomatic way, we briefly recall their construction: Fix a noncommutative probability space $(\mathcal{M}, \varphi)$, a von Neumann algebra $\mathcal{M}$ acting faithfully on the Hilbert space H together with a normal faithful state φ.

The spaces $L_p(\mathcal{M}, \varphi)$ are constructed as spaces of measurable operators not relative to $(\mathcal{M}, \varphi)$ but to a certain semifinite super von Neumann algebra of $\mathcal{M}$, the crossed product

$$\mathcal{R} := \mathcal{M} \rtimes_\sigma \mathbb{R}$$

of $\mathcal{M}$ by $\mathbb{R}$ with respect to the modular automorphism group $\sigma = (\sigma_t^\varphi)_{t \in \mathbb{R}}$. Recall that this is the von Neumann algebra of operators on $L_2(\mathbb{R}, H)$ generated by the operators $\pi(x), x \in \mathcal{M}$ and the operators $\lambda(s), s \in \mathbb{R}$ defined as follows: for any $\xi \in L_2(\mathbb{R}, H)$ and $t \in \mathbb{R}$

$$[(\pi x)\xi](t) := [\sigma_{-t}^\varphi x]\xi(t) \text{ and } [\lambda(s)\xi](t) := \xi(t-s).$$

Note that π is a faithful represetation of $\mathcal{M}$ on $L_2(\mathbb{R}, H)$, and therefore $\mathcal{M}$ may be identified with the subalgebra $\pi(\mathcal{M})$ in $\mathcal{R}$.

Let $(\theta_t(x))_{t \in \mathbb{R}}$ be the dual automorphism group on $\mathcal{R}$:

$$\theta_t(x) := W(t)xW(t)^* \text{ for all } t \in \mathbb{R}, x \in \mathcal{R},$$

where $W(t)(\xi)(s) := e^{-its}\xi(s)$, $\xi \in L_2(\mathbb{R}, H)$, $t, s \in \mathbb{R}$ is the so called phase shift. It is well know that the position of $\mathcal{M}$ in $\mathcal{R}$ is determined by this group in the following sense: For every $x \in \mathcal{R}$ we have

$$x \in \mathcal{M} \iff \theta_t(x) = x \text{ for all } t \in \mathbb{R}.$$

Moreover, the crossed product $\mathcal{R}$ is semifinite, and it therefore can be equipped with a unique normal, semifinite and faithful trace τ satisfying

$$\tau(\theta_t(x)) = e^{-t}\tau(x) \text{ for all } t \in \mathbb{R}, x \in \mathcal{R}.$$

All $*$-automorphisms $\theta_s, s \in \mathbb{R}$ extend to $*$-automorphisms on $L_0(\mathcal{R}, \tau)$, all τ-measurable operators on $L_2(\mathbb{R}, H)$ τ-affiliated with $\mathcal{R}$. Now for $1 \leq p \leq \infty$ the Haagerup noncommutative L_p-space

$$L_p(\mathcal{M}) = L_p(\mathcal{M}, \varphi)$$

is defined to be the space of all $x \in L_0(\mathcal{R}, \tau)$ such that

$$\theta_s(x) = e^{-\frac{s}{p}}x \text{ for all } s \in \mathbb{R}. \tag{3.45}$$

The spaces $L_p(\mathcal{M}, \varphi)$ are closed and selfadjoint linear subspaces of $L_0(\mathcal{R}, \tau)$; they are closed under left and right multiplication by elements from $\mathcal{M}$. If $x = u|x|$ is the polar decomposition of $x \in L_0(\mathcal{R}, \tau)$, then

$$x \in L_p(\mathcal{M}, \varphi) \iff u \in \mathcal{M} \text{ and } |x| \in L_p(\mathcal{M}, \varphi); \tag{3.46}$$

as a consequence the left and right supports of $x \in L_p(\mathcal{M}, \varphi)$ belong to $\mathcal{M}$.

It is not hard to see that $L_\infty(\mathcal{M}, \varphi)$ coincides with $\mathcal{M}$ (modulo the inclusion $\mathcal{M} \hookrightarrow \mathcal{R} \hookrightarrow L_0(\mathcal{R}, \tau)$). Moreover, it is possible to show that there is a linear homeomorphism

$$\mathcal{M}_* \to L_1(\mathcal{M}, \varphi), \ \psi \to a_\psi$$

preserving all the additional structure like conjugation, positivity, polar decomposition, and action on $\mathcal{M}$. Clearly, this mapping permits to transfer the norm on the predual $\mathcal{M}_*$ to a norm on $L_1(\mathcal{M}, \varphi)$ denoted by $\| \cdot \|_1$.

One of the key points in the whole construction is the distinguished trace functional

$$\mathrm{tr} : L_1(\mathcal{M}, \varphi) \to \mathbb{C}, \ \mathrm{tr}(a_\psi) = \psi(1);$$

consequently, $\|x\|_1 = \mathrm{tr}(|x|)$ for all $x \in L_1(\mathcal{M}, \varphi)$. For every positive operator $x \in L_0(\mathcal{R}, \tau)$ we have that

$$x \in L_p(\mathcal{M}, \varphi) \iff x^p \in L_1(\mathcal{M}, \varphi)$$

Hence together with the norm

$$\|x\|_p = \mathrm{tr}(|x|^p)^{1/p} = \| |x|^p \|_1^{1/p}, \ x \in L_p(\mathcal{M}, \varphi)$$

the vector space $L_p(\mathcal{M}, \varphi)$ forms a Banach space, and $(L_2(\mathcal{M}, \varphi), \| \cdot \|_2)$ is a Hilbert space with the inner product

$$(x \mid y) = \mathrm{tr}(y^* x).$$

It can be proved that the vector space topology on $L_p(\mathcal{M}, \varphi)$ associated with the norm coincides with the topology inherited by $L_0(\mathcal{R}, \tau)$. Hölder's inequality extends to Haagerup L_p-spaces: If $1/r = 1/p + 1/q$, then for all $x \in L_p(\mathcal{M}, \varphi)$, $y \in L_q(\mathcal{M}, \varphi)$ we have

$$xy \in L_r(\mathcal{M}, \varphi) \text{ and } \|xy\|_r \leq \|x\|_p \|y\|_q .$$

In particular, the bilinear form $(x, y) \to \mathrm{tr}(xy)$ defines a duality bracket between $L_p(\mathcal{M}, \varphi)$ and $L_{p'}(\mathcal{M}, \varphi)$ for which the duality relation

$$L_p(\mathcal{M}, \varphi)' = L_{p'}(\mathcal{M}, \varphi)$$

holds isometrically; note that for $x \in L_p(\mathcal{M}, \varphi)$ and $y \in L_{p'}(\mathcal{M}, \varphi)$ the trace property

$$\mathrm{tr}(xy) = \mathrm{tr}(yx) \tag{3.47}$$

holds. The state φ being a normal positive functional on $\mathcal{M}$ corresponds to a positive element in $L_1(\mathcal{M}, \varphi)$ – this element we call density of φ, and denote it by D:

$$\varphi(x) = \mathrm{tr}(xD) = \mathrm{tr}(Dx) = \mathrm{tr}(D^{\frac{1}{r}}xD^{\frac{1}{s}}), \quad x \in L_1(\mathcal{M}, \varphi), \ 1/p = 1/r + 1/s$$

The following lemma will be crucial – for a proof see e.g. [39, Lemma 1.2] or [41, Lemma 1.1].

Lemma 23. *For $1 \le p < \infty$ and $1 \le r, s \le \infty$ with $1/p = 1/r + 1/s$ the linear mapping*

$$\mathcal{M} \hookrightarrow L_p(\mathcal{M}, \varphi), \quad x \rightsquigarrow D^{\frac{1}{r}}xD^{\frac{1}{s}}$$

is injective and has dense range.

It is known that each $L_p(\mathcal{M}, \varphi)$ is independent of φ in the sense that two spaces $L_p(\mathcal{M}, \varphi_1)$ and $L_p(\mathcal{M}, \varphi_2)$ can be identified up to an isomorphism preserving the norm, the order and the modular structure. This is the reason why we sometimes will simply write $L_p(\mathcal{M})$. Finally, we remark that if φ is tracial, i.e. $\varphi(xx^*) = \varphi(x^*x)$ for $x \in \mathcal{M}$, then the noncommutative Haagerup L_p-space $L_p(\mathcal{M}, \varphi)$ coincides with the non-commutative L_p-space used in the preceding section.

References. See e.g. [24, 37, 48, 80, 90].

3.2.2 Maximal Inequalities in Haagerup L_p-Spaces

The definition of ℓ_∞- and c_0-valued noncommutative L_p-spaces extends from Sect. 3.1.2 verbatim to the Haagerup setting. Fix again a noncommutative probability space $(\mathcal{M}, \varphi)$, a von Neumann algebra $\mathcal{M}$ together with a normal, faithful state φ. Moreover, let I be a countable partially ordered index set and $1 \le p < \infty$. Then exactly as in Sect. 3.1.2 we define

$$L_p(\mathcal{M})[\ell_\infty] = L_p(\mathcal{M}, \varphi)[\ell_\infty(I)]$$

to be the space of all families $(x_n)_{n \in I}$ of operators in $L_p(\mathcal{M}, \varphi)$ which admit a factorization

$$x_n = ay_nb, \ n \in I$$

with $a, b \in L_{2p}(\mathcal{M}, \varphi)$ and a bounded sequence in (y_n) in $\mathcal{M}$. Put

$$\|(x_n)\|_{L_p[\ell_\infty]} := \inf \|a\|_{2p} \sup_n \|y_n\|_\infty \|b\|_{2p},$$

the infimum taken over all possible factorizations. Again we also consider column and row variants of this notion; define

$$L_p(\mathcal{M})[\ell_\infty^c] = L_p(\mathcal{M}, \varphi)[\ell_\infty^c(I)]$$

to be the space of all sequences $(x_n)_{n \in I}$ in $L_p(\mathcal{M}, \varphi)$ for which there are some $b \in L_p(\mathcal{M}, \varphi)$ and a bounded sequence (y_n) in $\mathcal{M}$ such that

$$x_n = y_n b, \ n \in I,$$

and put again

$$\|(x_n)\|_{L_p[\ell_\infty^c]} := \inf \sup_n \|y_n\|_\infty \|b\|_p.$$

Of course, the symbol $L_p(\mathcal{M})[\ell_\infty^r] = L_p(\mathcal{M}, \varphi)[\ell_\infty^r]$ then stands for all sequences (x_n) which allow a uniform factorization $x_n = a y_n$. Following the lines of Sect. 3.1.2 we define the c_0-valued noncommutative Haagerup L_p's: If in the three above definitions we replace the bounded sequences (y_n) in $\mathcal{M}$ by zero sequences (y_n) in $\mathcal{M}$, then we obtain the subspaces

$$L_p(\mathcal{M}, \varphi)[c_0(I)], \ L_p(\mathcal{M}, \varphi)[c_0^c(I)] \text{ and } L_p(\mathcal{M}, \varphi)[c_0^r(I)],$$

respectively.

By now it is clear that all these definitions are crucial for our purposes, and as we already pointed out in Sect. 3.1.2 they have their origin in Pisier [79] and Junge [39] (see also [42]). The following result is the nontracial analogue of Proposition 3 and Proposition 4.

Proposition 7.

(1) $L_p(\mathcal{M})[\ell_\infty(I)]$ is a Banach space, and $L_p(\mathcal{M})[\ell_\infty^c(I)]$ and $L_p(\mathcal{M})[\ell_\infty^r(I)]$ are quasi Banach spaces which are even normed whenever $p \geq 2$.

(2) $L_p(\mathcal{M})[c_0(I)]$ is an isometric subspace of $L_p(\mathcal{M})[\ell_\infty(I)]$, and the same result holds in the column and row case.

Proof. The proof of (1) is a one to one copy of the proof of Proposition 3; the only difference is to observe that the following modification of Lemma 17 holds: *Given positive operators* $c, c_1 \in L_p(\mathcal{M}, \varphi)$ *with* $c_1^2 \leq c^2$, *there is a contraction* $a_1 \in \mathcal{M}$ *satisfying* $c_1 = a_1 c$. *Moreover, if* $c_1, c_2 \in L_p(\mathcal{M}, \varphi)$ *and* $c := (c_1^2 + c_2^2)^{1/2}$, *then there are contractions* $a_1, a_2 \in \mathcal{M}$ *such that*

$$c_k = a_k c \text{ for } k = 1, 2 \text{ and } a_1^* a_1 + a_2^* a_2 = r(c^2).$$

Since $L_p(\mathcal{M}, \varphi) \subset L_0(\mathcal{R}, \tau)$, this result in fact is a simple consequence of Lemma 17: From this lemma (and its proof) we see that the operators we look for are the contractions $a_k = c_k c^{-1} p_k$ in $\mathcal{R}$ where $p_k \in \mathcal{R}$ is the range projection of c_k^2. It remains to observe that these two operators are in $\mathcal{M}$; indeed, we know from (3.46) that $p_k \in \mathcal{M}$. Hence from (3.45) we conclude for every $s \in \mathbb{R}$ that $p_k = \theta_s(p_k) = \theta_s(cc^{-1} p_k) = e^{-s/p} c \theta_s(c^{-1} p_k)$, and therefore $\theta_s(c^{-1} p_k) = e^{s/p} c^{-1} p_k$. But then we finally get that

$$\theta_s(a_k) = \theta_s(c_k) \theta_s(c^{-1} p_k) = e^{-s/p} c_k e^{s/p} c^{-1} p_k = a_k \in \mathcal{M}.$$

The above modification of Lemma 17 also shows that the proof of (2) follows from a word by word translation of the proof of Proposition 4.   □

3.2.3 Nontracial Extensions of Maximizing Matrices

The following extension theorem for $(2,2)$-maximizing matrices within Haagerup L_p-spaces is again crucial – the result is an analogue of Theorem 13 (commutative case), and of the Theorems 17 and 19 (noncommutative tracial case). As in the preceding section the following two results are stated with respect to a fixed noncommutative probability space $(\mathcal{M}, \varphi)$.

Theorem 24. *Let* $A = (a_{jk})$ *be a* $(2,2)$*-maximizing matrix. Then for each unconditionally convergent series* $\sum_k x_k$ *in* $L_p(\mathcal{M}, \varphi)$ *the following holds:*

(1) $\left(\sum_{k=0}^{\infty} a_{jk} x_k \right)_j \in L_p(\mathcal{M})[\ell_{\infty}]$

(2) $\left(\sum_{k=0}^{\infty} a_{jk} x_k \right)_j \in L_p(\mathcal{M})[\ell_{\infty}^c]$ *provided* $p \geq 2$

(3) *If* $p \leq 2$*, then* $\sum_k x_k$ *splits into a sum*

$$\sum_k x_k = \sum_k u_k + \sum_k v_k$$

of two unconditionally convergent series such that

$$\left(\sum_{k=0}^{\infty} a_{jk} u_k \right)_j \in L_p(\mathcal{M})[\ell_{\infty}^r] \text{ and } \left(\sum_{k=0}^{\infty} a_{jk} v_k \right)_j \in L_p(\mathcal{M})[\ell_{\infty}^c].$$

Proof. For the proof of (1) and (2) check again the proof of Theorem 17. For the proof of (3) analyze the proof of the row+column maximal Theorem 19. Note that for the proof of Lemma 19 for $L_p(\mathcal{M}, \varphi)$, $1 < p \leq 2$ again the Lust-Picard domination theorem from (3.24) is needed, now in the nontracial case (in order to check this nontracial Lust-Picard domination theorem either copy the proof from [55], or deduce it from the tracial one through Haagerup's reduction method elaborated in [26]). For the proof of Lemma 19 for $L_1(\mathcal{M}, \varphi)$ copy the argument given in the tracial case word by word.   □

In view of the fact that every $(2,2)$-maximizing matrix is $(1,\infty)$-maximizing (see again Theorem 5) the following analog of Theorem 18 extends the statements (1) and (2) of the preceding theorem.

Theorem 25. *Let* $A = (a_{jk})$ *be a* (p,q)*-maximizing matrix. Then for each* $\alpha \in \ell_q$ *and each weakly* q'*-summable sequence* (x_k) *in* $L_p(\mathcal{M}, \varphi)$

$$\left(\sum_{k=0}^{\infty} a_{jk} \alpha_k x_k \right)_j \in L_p(\mathcal{M}, \varphi)[\ell_{\infty}].$$

Moreover, ℓ_{∞} *can be replaced by* ℓ_{∞}^c *provided* $p \geq 2$*.*

For $p \leq 2$, $q = \infty$ and unconditionally summable sequences (x_k) we here could also prove a row+column version like in part (3) of Theorem 24 – but since for $p \leq 2$ by Theorem 5 a matrix A is (p, ∞)-maximizing if and only if it is $(2, 2)$-maximizing, such a statement is already contained in Theorem 24.

Proof. Basically, the proof runs along the lines of the proof of Theorem 18 – hence we here only sketch the main differences. We need part of Proposition 5 and then as a consequence Lemma 18 in the nontracial case.

The proof of the complex interpolation formula from Proposition 5 given in [42, Proposition 1.1] very much relies on two facts. The first one is the duality of $L_p(\mathcal{M}, \varphi)[\ell_\infty]$ and $L_{p'}(\mathcal{M}, \varphi)[\ell_1]$ which is proved in [39, Proposition 3.6] and holds true in the tracial as well as non tracial case (see also [42, Proposition 2.1,(iii)]). The second ingredient is complex interpolation of noncommutative L_p's which in the semifinite case works perfectly but is less comfortable in the nontracial case (see the discussion from [80, Sect. 2]).

But note that for the proof of Lemma 18 it is not necessary to know Proposition 5 in its full generality. In (3.14) and (3.15) we only need to know that complex interpolation of $L_1(\mathcal{M}, \varphi)[\ell_\infty]$ with $L_\infty(\mathcal{M}, \varphi)[\ell_\infty]$ and of $L_1(\mathcal{M}, \varphi)[\ell_\infty^c]$ with $L_2(\mathcal{M}, \varphi)[\ell_\infty^c]$ works properly, and in order to prove this we need (the above mentioned duality and) complex interpolation of Haagerup L_p's in the three cases $p = 1, 2, \infty$. This is possible: Every Haagerup L_p can be represented in form of a Kosaki L_p which by definition is the complex interpolation space $(\mathcal{M}, \mathcal{M}_*)_{1/p} = (L_\infty(\mathcal{M}, \varphi), L_1(\mathcal{M}, \varphi))_{1/p}$ (see e.g. [48, Definition 3.1] and again [80, Sect. 3.]); for the interpolation of $L_1(\mathcal{M}, \varphi)$ with $L_2(\mathcal{M}, \varphi)$ see [48, Remark 3.4].

All in all this allows to reproduce Lemma 18 for Haagerup L_p's, and as explained above there then is no problem left to copy the proof of Theorem 18 in order to obtain Theorem 25. $\square$

3.2.4   Almost Sure Convergence

In this section we show how to deduce from the preceding maximal theorems almost everywhere convergence theorems in noncommutative Haagerup L_p's – this is more complicated than in the tracial case. One of the reasons (another was given in the introduction of this section) is that the operators in $L_p(\mathcal{M}, \varphi)$ are no longer affiliated with $\mathcal{M}$ itself but affiliated with a larger von Neumann algebra, the crossed product with respect to the modular automorphism group.

Fix again some von Neumann algebra $\mathcal{M}$ of operators acting on the Hilbert space H, and a faithful and normal state φ on $\mathcal{M}$ with density D. We need to introduce appropriate analogues of almost everywhere convergence in Haagerup L_p's. There are several such generalizations. Here we adopt the notion of almost sure convergence, first introduced by Hensz and Jajte in [33] for sequences in $L_2(\mathcal{M}, \varphi)$ in order to study noncommutative Menchoff-Rademacher type theorems in such spaces (see [38, Definition 1.1.1.]). We now extend this notion to almost sure, bilateral almost sure and row+column almost sure convergence in Haagerup L_p-spaces; see also [39] and [42]. Given a sequence (y_n) and some operator y in $L_p(\mathcal{M}, \varphi)$, we say that:

- (y_n) converges to y φ-almost surely whenever for every $\varepsilon > 0$ there is a projection $s \in \mathcal{M}_{\text{proj}}$ together with a matrix (a_{nk}) in $\mathcal{M}$ such that $\varphi(1 - s) < \varepsilon$, $y_n - y = \sum_k a_{nk} D^{1/p}$ for all n, and $\lim_n \| \sum_k a_{nk} s \|_\infty = 0$; here the two series converge in $L_p(\mathcal{M}, \varphi)$ and in $\mathcal{M}$, respectively.
- (y_n) converges to y bilaterally φ-almost surely whenever for every $\varepsilon > 0$ there is a projection $s \in \mathcal{M}_{\text{proj}}$ together with a matrix (a_{nk}) in $\mathcal{M}$ such that $\varphi(1 - s) < \varepsilon$, $y_n - y = \sum_k D^{1/2p} a_{nk} D^{1/2p}$ for all n, and $\lim_n \| \sum_k s a_{nk} s \|_\infty = 0$.
- (y_n) converges row+column φ-almost surely provided the sequence (y_n) decomposes into a sum $(a_n) + (b_n)$ of two sequences where (a_n) and (b_n^*) both are converging φ-almost surely.

The following device is a perfect nontracial analog of Proposition 6, and it is again crucial for our purposes.

Proposition 8. *Assume that the sequence (x_n) from $L_p(\mathcal{M}, \varphi)$ satisfies*

$$(x_n - x_m)_{nm} \in L_p(\mathcal{M})[c_0(\mathbb{N}_0^2)]. \tag{3.48}$$

Then (x_n) converges to some x in $L_p(\mathcal{M})$, and

$$x = \lim_n x_n \ \text{bilaterally } \varphi\text{-almost surely};$$

if in (3.48) we replace c_0 by c_0^c, then bilateral φ-almost sure convergence can be replaced by φ-almost sure convergence, and if we have c_0^{r+c} instead of c_0, then it is possible to replace bilateral φ-almost sure convergence by row+column φ-almost sure convergence.

Part of this proposition was first observed within our study of noncommutative coefficient tests and later published in [42, Lemma 7.10] (the main part of the proof of Proposition 8 is taken from this reference). The argument that (x_n) in fact converges (bilaterally) φ-almost surely to the limit x of the L_p-Cauchy sequence (x_n) is a modification of the proof of [34, Proposition 3.2] due to Hensz, Jajte and Paszkiewicz.

We will need several lemmata in order to prove this proposition – the proof will be given after Lemma 26. The first one is a Chebychev type inequality of Goldstein [19]. For us this result will turn out to be fundamental – as fundamental as it is for the work of Hensz, Jajte, and their coauthors for all their study on strong limit theorems in noncommutative probabilty (see [37, Sect. 2.2, Corollary 2.2.13] and [38, Sect. 2.1, Theorem 2.2.1]).

Lemma 24. *Let (x_n) be a sequence of positive operators in $\mathcal{M}$ and (ε_n) a sequence of positive scalars such that $\sum_{n=1}^\infty \frac{\varphi(x_n)}{\varepsilon_n} < \frac{1}{2}$. Then there is a projection s in $\mathcal{M}$ such that*

$$\varphi(1-s) \leq 2 \sum_{n=1}^{\infty} \frac{\varphi(x_n)}{\varepsilon_n} \text{ and } \|sx_n s\|_{\infty} \leq 2\varepsilon_n , \, n \in \mathbb{N}.$$

For the second lemma we need a straight forward reformulation of the above defintions of almost sure convergence. For $y \in L_p(\mathcal{M}, \varphi)$ and s a projection in $\mathcal{M}$ define

$$\|y\|_{[s]}^c = \inf \left\| \sum_{k=0}^{\infty} y_k s \right\|_{\infty},$$

where the infimum is taken over all sequences (y_k) in $\mathcal{M}$ such that $\sum_{k=0}^{\infty} y_k s$ converges in $\mathcal{M}$ and $y = \sum_{k=0}^{\infty} y_k D^{1/p}$ converges in $L_p(\mathcal{M}, \varphi)$ (with the usual convention $\inf \emptyset = \infty$). In the symmetric case we define similarly:

$$\|y\|_{[s]} = \inf \left\| \sum_{k=0}^{\infty} s y_k s \right\|_{\infty},$$

the infimum now taken over all sequences (y_k) in $\mathcal{M}$ such that $\sum_{k=0}^{\infty} s y_k s$ converges in $\mathcal{M}$ and $y = \sum_{k=0}^{\infty} D^{1/2p} y_k D^{1/2p}$ converges in $L_p(\mathcal{M}, \varphi)$. With these notions it then can be seen easily that a sequence (y_n) in $L_p(\mathcal{M}, \varphi)$ converges to $y \in L_p(\mathcal{M}, \varphi)$

- φ-almost surely if for each ε there is a projection s in $\mathcal{M}$ such that $\varphi(1-s) \leq \varepsilon$ and $\|y_n - y\|_{[s]}^c \to 0$ as $n \to \infty$,
- and bilaterally φ-almost surely whenever we in the preceding definition of almost sure convergence only have $\|y_n - y\|_{[s]} \to 0$.

The following lemma will be useful; its proof modifies the argument from [34, Lemma 3.1].

Lemma 25. *Let $\sum_n z_n$ be a convergent series in $L_p(\mathcal{M}, \varphi)$ and s a projection in $\mathcal{M}$. Then*

$$\left\| \sum_{n=1}^{\infty} z_n \right\|_{[s]} \leq \sum_{n=1}^{\infty} \|z_n\|_{[s]},$$

and the same inequality holds for $\|\cdot\|_{[s]}^c$ instead of $\|\cdot\|_{[s]}$.

Proof. Put $z = \sum_{n=1}^{\infty} z_n$, and assume without loss of generality that

$$\sum_{n=1}^{\infty} \|z_n\|_{[s]} < \infty. \tag{3.49}$$

Fix some $\varepsilon > 0$, and define $\varepsilon_n = \varepsilon/2^n$ and $\varepsilon_{nm} = \varepsilon/2^{n+m}$. Since all $\|z_n\|_{[s]} < \infty$, we by definition of $\|\cdot\|_{[s]}$ may choose a matrix $(z_{n,k})_{n,k\in\mathbb{N}}$ in $\mathcal{M}$ such that for all n, m we have

$$\left\| z_n - \sum_{k=1}^{m} D^{1/2p} z_{n,k} D^{1/2p} \right\|_p < \varepsilon_{nm} \tag{3.50}$$

$$\|z_{n,m+1}\|_{\infty} < \varepsilon_{nm} \tag{3.51}$$

$$\left\| \sum_{k=1}^{m} sz_{n,k}s \right\|_{\infty} < \|z_n\|_{[s]} + \varepsilon_n. \tag{3.52}$$

We show for $x_n := \sum_{k,\ell=1}^{n} z_{k,\ell} \in \mathcal{M}, n \in \mathbb{N}$ that $\lim_n \|z - D^{1/2p}x_n D^{1/2p}\|_p = 0$; in fact, by (3.50)

$$\|z - D^{1/2p}x_n D^{1/2p}\|_p \leq \sum_{k=1}^{n} \left\| z_k - \sum_{\ell=1}^{n} D^{1/2p}z_{k,\ell}D^{1/2p} \right\|_p + \left\| \sum_{k=n+1}^{\infty} z_k \right\|_p$$

$$\leq \sum_{k=1}^{n} \varepsilon_{kn} + \left\| \sum_{k=n+1}^{\infty} z_k \right\|_p \to 0, \; n \to \infty.$$

Now note that $(sx_n s)_n$ converges in $\mathcal{M}$; indeed, we deduce from (3.51) and (3.52) that

$$\|s(x_{n+1} - x_n)s\|_{\infty} \leq \sum_{k=1}^{n} \|sz_{k,n+1}s\|_{\infty} + \left\| \sum_{\ell=1}^{n+1} sz_{n+1,\ell}s \right\|_{\infty}$$

$$< \sum_{k=1}^{n} \varepsilon_{k,n+1} + \|z_{n+1}\|_{[s]} + \varepsilon_{n+1} < 2\varepsilon_{n+1} + \|z_{n+1}\|_{[s]},$$

hence we see as a consequence of (3.49) that $(sx_n s)_n$ is a Cauchy sequence in $\mathcal{M}$. Finally, (3.52) implies that

$$\|sx_n s\|_{\infty} \leq \sum_{k=1}^{n} \left\| \sum_{\ell=1}^{n} sz_{k,\ell}s \right\|_{\infty} < \sum_{k=1}^{n} \|z_k\|_{[s]} + \varepsilon_k < \sum_{k=1}^{n} \|z_k\|_{[s]} + \varepsilon$$

which clearly gives the desired inequality. The column case follows similarly. □

As an easy consequence of the preceding lemma we now deduce the following noncommutative Cauchy condition for almost sure convergence which in the L_2-case was given in [34, Proposition 3.2]; again our proof adapts the arguments from this reference.

Lemma 26. *Let (y_n) be a convergent sequence in $L_p(\mathcal{M}, \varphi)$ with limit y, and such that for each $\varepsilon > 0$ there is a projection s in $\mathcal{M}$ with $\varphi(1 - s) < \varepsilon$ and $\|y_n - y_m\|_{[s]} \to 0$ as $n, m \to \infty$. Then*

$$y = \lim_n y_n \text{ bilaterally } \varphi\text{-almost surely};$$

moreover, we here may replace bilateral φ-almost sure convergence by φ-almost sure convergence provided we change in the hypothesis $\|\cdot\|_{[s]}$ by $\|\cdot\|_{[s]}^c$.

Proof. Fix $\varepsilon > 0$. We show that there is some m_0 such that for all $n > m_0$ we have $\|y - y_n\|_{[s]} < 2\varepsilon$. According to the assumption there is a projection s in $\mathcal{M}$ with $\varphi(1 - s) < \varepsilon$ and a sequence of indices $m_0 < m_1 < \ldots$ such that for all k

$$\|y_n - y_m\|_{[s]} < \varepsilon/2^k \text{ for } n, m \geq m_k. \tag{3.53}$$

Take now some $n > m_0$, choose k such that $n < m_k$, and define

$$z_0 = y_{m_k} - y_n$$
$$z_j = y_{m_{k+j}} - y_{m_{k+j-1}} \text{ for } j = 1, 2, \ldots.$$

Then the series $\sum_{j=0}^{\infty} z_j$ in $L_p(\mathcal{M})$ converges, and by assumption its L_p-limit is given by $y - y_n$. Moreover, by (3.53) we see that

$$\|z_0\|_{[s]} < \varepsilon, \|z_1\|_{[s]} < \varepsilon/2^k, \ldots, \|z_j\|_{[s]} < \varepsilon/2^{k+j-1}, \ldots$$

and therefore $\sum_{j=0}^{\infty} \|z_j\|_{[s]} < 2\varepsilon$. But then Lemma 25 gives $\|y - y_n\|_{[s]} < 2\varepsilon$, the conclusion. The column case follows similarly.   □

Finally, we are in the position to give the *Proof of Proposition* 8: Again we only show the symmetric case – the column case follows similarly. We assume that the sequence $(x_n)_n$ from $L_p(\mathcal{M}, \varphi)$ satisfies

$$(x_n - x_m)_{nm} \in L_p(\mathcal{M})[c_0(\mathbb{N}_0^2)], \tag{3.54}$$

and show that for each $\varepsilon > 0$ there is a projection s in $\mathcal{M}$ with $\varphi(1 - s) < \varepsilon$ and

$$\|x_n - x_m\|_{[s]} \to 0 \text{ as } n, m \to \infty.$$

Since the assumption (3.54) in particular implies that the sequence $(x_n - x_m)_{nm}$ is a Cauchy sequence in $L_p(\mathcal{M})$, the conclusion then follows from Lemma 26. By definition there are $a, b \in L_p(\mathcal{M})$ and a matrix $(y_{nm})_{nm}$ in $\mathcal{M}$ such that

$$x_n - x_m = a y_{nm} b \text{ and } \|a\|_{2p} < 1, \|b\|_{2p} < 1, \lim_{n,m \to \infty} \|y_{nm}\|_\infty = 0.$$

From Lemma 23 we deduce that there are $a_k \in \mathcal{M}$ and $b_k \in \mathcal{M}$ such that

$$a = \sum_k D^{\frac{1}{2p}} a_k \text{ and } \|D^{\frac{1}{2p}} a_k\|_{2p} < 2^{-k}$$

$$b = \sum_k b_k D^{\frac{1}{2p}} \text{ and } \|b_k D^{\frac{1}{2p}}\|_{2p} < 2^{-k}.$$

Then

$$x_n - x_m = \sum_{j,k} D^{\frac{1}{2p}} a_j y_{nm} b_k D^{\frac{1}{2p}},$$

where the series converges absolutely in $L_p(\mathcal{M})$. It suffices to check that for each $\varepsilon > 0$ there is a projection s in $\mathcal{M}$ such that $\varphi(1-s) < \varepsilon$ and

$$\lim_{n,m \to \infty} \left\| \sum_{j,k} s a_j y_{nm} b_k s \right\|_{\infty} = 0.$$

Fix some $0 < \varepsilon < 1/2$. By Hölder's inequality and the trace propery (3.47) we obtain

$$\varphi(a_k a_k^*) = \text{tr}(a_k a_k^* D^{\frac{1}{2p}} D^{\frac{1}{p'}} D^{\frac{1}{2p}})$$
$$= \text{tr}(D^{\frac{1}{2p}} a_k a_k^* D^{\frac{1}{2p}} D^{\frac{1}{p'}}) \le \left\| D^{\frac{1}{2p}} a_k a_k^* D^{\frac{1}{2p}} \right\|_p < 2^{-2k}, \tag{3.55}$$

and similarly $\varphi(b_k^* b_k) < 2^{-2k}$. From Goldstein's Lemma 24 (put $\varepsilon_n = \frac{4}{2^n \varepsilon}$) we then know that there is a projection $s \in \mathcal{M}$ such that $\varphi(1-s) < \varepsilon$ and for each k

$$\max\{\|s a_k a_k^* s\|_\infty, \|s b_k^* b_k s\|_\infty\} \le 8\varepsilon^{-1} 2^{-k}.$$

As a consequence we get that for each n, m

$$\left\| \sum_{j,k} s a_j y_{nm} b_k s \right\|_\infty \le \sum_{j,k} \|s a_j y_{nm} b_k s\|_\infty$$

$$\le \|y_{nm}\|_\infty \sum_k \|s a_k a_k^* s\|_\infty^{\frac{1}{2}} \sum_k \|s b_k^* b_k s\|_\infty^{\frac{1}{2}}$$

$$\le \|y_{nm}\|_\infty 8\varepsilon^{-1} \left(\frac{\sqrt{2}}{\sqrt{2}-1} \right)^2 \to 0 \text{ as } n, m \to \infty,$$

the conclusion of Proposition 8.

3.2.5 Coefficient Tests in Haagerup L_p-Spaces

As always in this third chapter, we fix some von Neumann algebra $\mathcal{M}$ of operators acting on the Hilbert space H, and a faithful and normal state φ on $\mathcal{M}$. This section is the nontracial analogue of Sect. 3.1.6. In Sect. 3.2.2 we showed how to extend maximizing matrices within Haagerup L_p-spaces. Following our general program we apply these results to concrete summation processes and deduce noncommutative coefficient tests on almost sure summation of unconditionally convergent series in such spaces.

Again we need a substitute of Lemma 22 for almost sure instead of almost uniform convergence, and although the result is very similar to Lemma 22 we for the sake of completeness present it here in full detail.

Lemma 27. *Let $A = (a_{jk})$ be an infinite matrix which converges in each column and satisfies $\|A\|_\infty < \infty$. Let $1 \leq q \leq \infty$, and assume that*

$$\left(\sum_{k=0}^{\infty} a_{jk}\alpha_k x_k \right)_j \in L_p(\mathcal{M}, \varphi)[\ell_\infty] \tag{3.56}$$

for every sequence $(\alpha_k) \in \ell_q$ and every weakly q'-summable sequence (x_k) in $L_p(\mathcal{M}, \varphi)$ (in the case $q = \infty$ we only consider unconditionally summable sequences). Then for every such (α_k) and (x_k) the sequence $\left(\sum_{k=0}^{\infty} a_{jk}\alpha_k x_k \right)_j$ converges to some $s \in L_p(\mathcal{M}, \varphi)$, and

(1) $s = \lim_j \sum_{k=0}^{\infty} a_{jk}\alpha_k x_k$ bilaterally φ-almost surely
(2) If the assumption in (3.56) holds for ℓ_∞^c instead of ℓ_∞, then the convergence in (1) is even φ-almost sure.
(3) If the assumption in (3.56) holds for ℓ_∞^{r+c} instead of ℓ_∞, then the convergence in (1) is row+column φ-almost sure.

Proof. The proof again follows from a straight forward analysis of the proof of Lemma 7; for (1) show first like in (2.26) that for every (α_k) and (x_k) as in the statement we have

$$\left(\sum_{k=0}^{\infty} a_{ik}\alpha_k x_k - \sum_{k=0}^{\infty} a_{jk}\alpha_k x_k \right)_{(i,j)} \in L_p(\mathcal{M})[c_0(\mathbb{N}_0^2)], \tag{3.57}$$

and then deduce the conclusion from Proposition 8. The statements (2) and (3) follow similarly.   □

We start our study of coefficient tests in Haagerup L_p's with a nontracial variant of Theorem 20 which collects many of the results proved so far and is one of our main contributions on classical summation in noncommutative Haagerup L_p's.

Theorem 26. *Assume that S is a summation method and ω a Weyl sequence with the additional property that for each orthonormal series $\sum_k \alpha_k x_k$ in $L_2(\mu)$ we have*

$$\sup_j \left| \sum_{k=0}^{\infty} s_{jk} \sum_{\ell=0}^{k} \frac{\alpha_\ell}{\omega_\ell} x_\ell \right| \in L_2(\mu).$$

Then for each unconditionally convergent series $\sum_k x_k$ in $L_p(\mathcal{M}, \varphi)$ the following two statements hold:

(1) $\left(\sum_{k=0}^{\infty} s_{jk} \sum_{\ell=0}^{k} \frac{x_\ell}{\omega_\ell} \right)_j \in L_p(\mathcal{M})[\ell_\infty]$
(2) $\sum_{k=0}^{\infty} \frac{x_k}{\omega_k} = \lim_j \sum_{k=0}^{\infty} s_{jk} \sum_{\ell=0}^{k} \frac{x_\ell}{\omega_\ell}$ bilaterally φ-almost surely.

Moreover:

(3) If $2 \le p$, then the sequence in (1) belongs to $L_p(\mathcal{M})[\ell_\infty^c]$, and the convergence in (2) is even φ-almost sure.

(4) If $p \le 2$, then the series $\sum_k x_k$ splits into a sum

$$\sum_k x_k = \sum_k u_k + \sum_k v_k$$

of two unconditionally convergent series for which

$$\left(\sum_{k=0}^{\infty} s_{jk} \sum_{\ell=0}^{k} \frac{u_\ell}{\omega_\ell} \right)_j \in L_p(\mathcal{M})[\ell_\infty^r] \ , \ \left(\sum_{k=0}^{\infty} s_{jk} \sum_{\ell=0}^{k} \frac{v_\ell}{\omega_\ell} \right)_j \in L_p(\mathcal{M})[\ell_\infty^c].$$

In this case, the convergence in (2) is row+column φ-almost sure.

Proof. Conclude again from the assumption on S and Theorem 1 that $S\Sigma D_{1/\omega}$ is $(2,2)$-maximizing. Then all statements follow from Theorem 24 and Lemma 27.

$\square$

As in the preceding chapter we apply our results to ordinary summation, Riesz, Cesàro, and Abel summation – compare with Corollary 6 (the commutative case for vector-valued Banach function spaces) and Corollary 10 (the noncommutative tracial case for symmetric spaces of operators).

Corollary 14. *Let $\sum_k x_k$ be an unconditionally convergent series in $L_p(\mathcal{M}, \varphi)$. Then the following results hold:*

Maximal theorems:

(1) $\left(\sum_{k=0}^{j} \frac{x_k}{\log k} \right)_j \in L_p(\mathcal{M})[\ell_\infty]$

(2) $\left(\sum_{k=0}^{j} \frac{\lambda_{k+1} - \lambda_k}{\lambda_{j+1}} \sum_{\ell \le k} \frac{x_\ell}{\log \log \lambda_\ell} \right)_j \in L_p(\mathcal{M})[\ell_\infty]$ *for every strictly increasing, unbounded and positive sequence (λ_k) of scalars*

(3) $\left(\sum_{k=0}^{j} \frac{A_{j-k}^{r-1}}{A_j^r} \sum_{\ell \le k} \frac{x_\ell}{\log \log \ell} \right)_j \in L_p(\mathcal{M})[\ell_\infty]$ *for every $r > 0$*

(4) $\left(\sum_{k=0}^{\infty} \rho_j^k \frac{x_k}{\log \log k} \right)_j \in L_p(\mathcal{M})[\ell_\infty]$ *for every positive strictly increasing sequence (ρ_j) converging to 1*

(5) *If $p \ge 2$, then ℓ_∞ in (1)–(4) may be replaced by ℓ_∞^c, and if $p \le 2$, then in all four statements there is a decomposition*

$$\sum_k s_{jk} \sum_{\ell \le k} \frac{x_\ell}{\omega_\ell} = \sum_k s_{jk} \sum_{\ell \le k} \frac{u_\ell}{\omega_\ell} + \sum_k s_{jk} \sum_{\ell \le k} \frac{v_\ell}{\omega_\ell},$$

where the first summand belongs to $L_p(\mathcal{M})[\ell_\infty^r]$ and the second one to $L_p(\mathcal{M})[\ell_\infty^c]$.

Almost sure convergence:

(6) Moreover, in (1)–(4) we have that

$$\sum_{k=0}^{\infty} \frac{x_k}{\omega_k} = \lim_j \sum_{k=0}^{\infty} s_{jk} \sum_{\ell \leq k} \frac{x_\ell}{\omega_\ell} \quad \text{bilaterally } \varphi\text{-almost surely;}$$

if $p \geq 2$, then this convergence is φ-almost sure, and if $p \leq 2$, then it is row+column φ-almost sure.

The following noncommutative Menchoff-Rademacher type theorem which was first published by Hensz and Jajte in [33, Theorem 4.1] (see also [38, Theorem 5.2.1] and [36]), is an immediate consequence.

Corollary 15. *If (ξ_k) is an orthogonal sequence in the Hilbert space $L_2(\mathcal{M}, \varphi)$ such that*

$$\sum_k \log^2 k \, \|\xi_k\|_2^2 < \infty.$$

Then

$$\sum_{k=0}^{\infty} \xi_k = \lim_n \sum_{k=0}^{n} \xi_k \quad \varphi\text{-almost surely.}$$

Proof. Obviously $\sum_k \log k \, \|\xi_k\|_2 \frac{\xi_k}{\|\xi_k\|_2}$ is an orthonormal series in $L_2(\mathcal{M}, \varphi)$ which in particular is unconditionally convergent. Then the conclusion is a very particular case of Corollary 14, (6) (applied to ordinary summation and the Weyl sequence $\omega_k = \log k$ with $p = 2$). □

In [20, Theorem 5.2, 5.3] Goldstein and Łuczak proved a Menchoff-Rademacher type theorem for orthogonal sequences of selfadjoint operators in $L_2(\mathcal{M}, \varphi)$. There is a long series [27, 28, 29, 31, 30, 32] of papers studying noncommutative coefficient tests in $L_2(\mathcal{M}, \varphi)$ due to Hensz, and it seems that most of these results are extended and complemented by Corollary 14 – let us illustrate this with three examples:

- In [32, Sect. 4.1.2] the following result for *non*orthogonal sequences is proved: Let (ξ_k) be a sequence in $L_2(\mathcal{M}, \varphi)$ such that

$$\sum_{k,\ell} \log(k) \log(\ell) \left| \mathrm{tr}(\xi_k \xi_\ell) \right| < \infty. \tag{3.58}$$

Then $\sum_k \xi_k$ converges φ-almost surely; indeed, this is a simple consequence of our results: Assume that (ξ_k) satisfies (3.58). Then for every scalar sequence (ε_k) with $|\varepsilon_k| \leq 1$ we have

$$\left\| \sum_k \varepsilon_k \log(k) \xi_k \right\|_2^2 = \left| \sum_{k,\ell} \varepsilon_k \varepsilon_\ell \log(k) \log(\ell) \, \mathrm{tr}(\xi_k \xi_\ell) \right|$$

$$\leq \sum_{k,\ell} \log(k) \log(\ell) \left| \mathrm{tr}(\xi_k \xi_\ell) \right| < \infty,$$

hence the series $\sum_k \log(k)\,\xi_k$ is unconditionally convergent in $L_2(\mathcal{M},\varphi)$. But in this situation Corollary 14, (6) (ordinary summation, $\omega_k = \log k$, $p = 2$) yields the result of Hensz.

- A sequence (ξ_k) in $L_2(\mathcal{M},\varphi)$ is said to be quasi-orthogonal (see e.g. [38, Definition 5.2.4]) whenever there is a sequence $(\rho(k))$ of positive numbers such that

$$|(\xi_k|\xi_\ell)| \le \rho(|k-\ell|)\|\xi_k\|_2\|\xi_\ell\|_2 \text{ for all } k,\ell \text{ and } \sum_k \rho(k) < \infty. \qquad (3.59)$$

It can be seen easily that every sequence satisfying the two conditions (3.59) and $\sum_k \log^2(k)\|\xi_k\|_2^2 < \infty$ fulfills (3.58) (see [32, Sect. 4.1.7]); this shows that Corollary 14 also covers and extends most of the noncommutative Menchoff-Rademacher type results for quasi-orthogonal sequences due to Hensz; see e.g. [38, Theorem 5.2.5].

- In [27, Sect. 4] (see also [32, Sect. 4.2] and [38, Theorem 5.2.8]) Hensz proves the following noncommutative coefficient test for Cesàro summation: for every orthogonal sequence in $L_2(\mathcal{M},\varphi)$ for which $\sum_k \log(k)\|\xi_k\|_2^2 < \infty$ we have that

$$\sum_{k=0}^{\infty} \xi_k = \lim_n \frac{1}{n+1} \sum_{k=0}^{n} \sum_{\ell=0}^{k} \xi_\ell \quad \varphi\text{-almost surely.} \qquad (3.60)$$

This is a noncommutative extension of a classical result due to Weyl (weaker then the optimal result proved later by Kaczmarz and Menchoff, see again (1.7) and (1.8) from the introduction). Corollary 14 (statement (6) for Cesàro summation) shows that this is even true under the weaker condition $\sum_k (\log\log k)^2 \|\xi_k\|_2^2 < \infty$ (then $\sum_k \log\log(k)\,\xi_k = \sum_k \log\log(k)\|\xi_k\|_2 \frac{\xi_k}{\|\xi_k\|_2}$ is an orthonormal series, hence it is uncoditionally convergent in $L_2(\mathcal{M},\varphi)$ and therefore division through $\log\log k$ again gives (3.60)). The proof of this optimal noncommutative extension of the Kaczmarz-Menchoff theorem was left open in the work of Hensz and Jajte.

We can also give nontracial variants of Theorem 21 (see Theorem 14 and Corollary 7 for the corresponding results in vector-valued function spaces) – again, under restrictions on the series the logarithmic terms in Corollary 14 are superfluous.

Theorem 27. *Let S be a summation method, and $1 \le q < p < \infty$. Then for each $\alpha \in \ell_q$ and each weakly q'-summable sequence (x_k) in $L_p(\mathcal{M},\varphi)$ we have that*

(1) $\left(\sum_{k=0}^{\infty} s_{jk} \sum_{\ell \le k} \alpha_\ell x_\ell\right)_j \in L_p(\mathcal{M})[\ell_\infty]$

(2) $\sum_{k=0}^{\infty} \alpha_k x_k = \lim_j \sum_{k=0}^{\infty} s_{jk} \sum_{\ell \le k} \alpha_\ell x_\ell$ *bilaterally φ-almost surely;*

if $2 \le p$, then here ℓ_∞ may be replaced by ℓ_∞^c, and bilaterally φ-almost sure by φ-almost sure convergence.

Proof. As in the proof of Theorem 21 we first apply Theorem 2 to Theorem 26, and then deduce the result as an immediate consequence of Lemma 27.   □

Again this result can be applied to noncommutative Cesàro, Riesz, and Abel summation (see Corollary 11 for the semifinite and Corollary 7 for the commutative case).

Like in Sect. 3.1.6 we finish with a nontracial variant of Corollary 12; since $L_p(\mathcal{M}, \varphi)$ has cotype $\max\{p, 2\}$ (see [16, Sect. 3] for the tracial case, and in the nontracial case use the fact that each $L_p(\mathcal{M}, \varphi)$ by Haagerup's reduction method from [26] is a complemented subspace of an ultraproduct of noncommutative L_p's constructed over von Neumann algebras with trace), the proof follows as in Corollary 8 (the commutative case).

Corollary 16. *Let $\sum_k x_k$ be an unconditionally convergent series in $L_p(\mathcal{M}, \varphi)$, and let x be its sum. Then there is a Riesz matrix $R^\lambda = (r_{jk}^\lambda)$ such that*

$$\left(\sum_{k=0}^{\infty} r_{jk}^\lambda \sum_{\ell \leq k} x_\ell \right)_j \in L_p(\mathcal{M})[\ell_\infty]$$

and

$$\lim_j \sum_{k=0}^{\infty} r_{jk}^\lambda \sum_{\ell \leq k} x_\ell = x \quad \text{bilaterally } \varphi\text{-almost surely}. \tag{3.61}$$

Moreover, if $2 \leq p$, then ℓ_∞ may be replaced by ℓ_∞^c and the convergence is φ-almost sure. If $p \leq 2$, then R^λ may be chosen in such a way that the equality in (3.61) holds row+column almost surely. Finally, in both cases corresponding maximal inequalities hold.

3.2.6 Laws of Large Numbers in Haagerup L_p-Spaces

This section forms a nontracial variant of Sect. 3.1.7 (the commutative results were given in Sect. 2.3.2). We again fix a noncommutative probability space $(\mathcal{M}, \varphi)$, a von Neumann algebra $\mathcal{M}$ of operators acting on the Hilbert spaces H together with a faithful and normal state φ.

The following result is taken from Jajte [38, Sect. 5.2.3]: *Let (x_k) be an orthogonal sequence (x_k) in $L_2(\mathcal{M}, \varphi)$ such that $\sum_k \frac{\log^2 k}{k^2} \|x_k\|_2^2 < \infty$. Then*

$$\lim_j \frac{1}{j+1} \sum_{k=0}^{j} x_k = 0 \quad \varphi\text{-almost surely}. \tag{3.62}$$

We will see that in view of our concrete examples from Sect. 2.2.4 the following theorem is a strong extension of this result and a nontracial counterpart of Theorem 22.

Theorem 28. *Let S be a lower triangular summation method. Assume that ω is an increasing sequence of positive scalars such that for each orthogonal sequence (x_k) in $L_2(\mu)$ with $\sum_k \frac{\omega_k^2}{k^2}\|x_k\|_2^2 < \infty$ we have*

$$\sup_j \left| \frac{1}{j+1} \sum_{k=0}^{j} s_{jk} \sum_{\ell=0}^{k} x_\ell \right| \in L_2(\mu).$$

Then for each unconditionally convergent series $\sum_k \frac{\omega_k}{k} x_k$ in $L_p(\mathcal{M},\varphi)$ the following two statements hold:

(1) $\left(\frac{1}{j+1} \sum_{k=0}^{j} s_{jk} \sum_{\ell=0}^{k} x_\ell \right)_j \in L_p(\mathcal{M})[\ell_\infty]$

(2) $\lim_j \frac{1}{j+1} \sum_{k=0}^{j} s_{jk} \sum_{\ell=0}^{k} x_\ell = 0$ *bilaterally φ-almost surely.*

Moreover:

(3) If $p \geq 2$, then the sequence in (1) is in $L_p(\mathcal{M})[\ell_\infty^c]$, and the series in (2) converges even φ-almost surely.

(4) If $p \leq 2$, then there is a decomposition

$$\frac{1}{j+1} \sum_k s_{jk} \sum_{\ell \leq k} x_\ell = \frac{1}{j+1} \sum_k s_{jk} \sum_{\ell \leq k} u_\ell + \frac{1}{j+1} \sum_k s_{jk} \sum_{\ell \leq k} v_\ell,$$

where the first summand belongs to $L_p(\mathcal{M})[\ell_\infty^r]$ and the second one to $L_p(\mathcal{M})[\ell_\infty^c]$. Moreover, the convergence in (2) is even row+column φ-almost sure.

Proof. We repeat the proof of Theorem 22 word by word in our nontracial setting – the only difference is that we now use Theorem 24 instead of Theorem 17 and Theorem 19 in order to deduce the maximal theorems, and Lemma 27 instead of Lemma 22 to deduce the convergence results.   □

If in a first step we apply the Menchoff-Rademacher Theorem 6 to Lemma 14, and in a second step use Theorem 28, then we get a far reaching extension of the result of Hensz and Jajte from (3.62); if we use the Theorems 8, 10, or 12 we obtain variants for Riesz, Cesàro or Abel summation. All these results involve certain Weyl sequences, but Hensz proved in [29] (see also Jajte [38, Sect. 5.2.13]) that in the case of Cesàro summation no logarithmic term is needed: if $\sum_k \frac{\|x_k\|_2^2}{k^2} < \infty$, then

$$\lim_j \frac{1}{j+1} \sum_{k=0}^{j} (1 - \frac{k}{j+1}) x_k = \lim_j \frac{1}{j+1} \sum_{k=0}^{j} \frac{1}{j+1} \sum_{\ell \leq k} x_\ell = 0 \ \ \varphi\text{-almost surely}.$$

As in Theorem 23 (tracial case) we obtain a similar result for Cesàro summation of order $r > 0$ – no log-terms are involved (as mentioned earlier these are noncommutative variants of results on orthonormal series in $L_2(\mu)$ due to Moricz [63]).

Theorem 29. *Assume that $\sum_k \frac{x_k}{k}$ is an unconditionally convergent series in $L_p(\mathcal{M}, \varphi)$ and $r > 0$.*

Maximal theorems:

(1)   We have

$$\left(\frac{1}{j+1} \sum_{k=0}^{j} \frac{A_{j-k}^{r-1}}{A_j^r} \sum_{\ell \leq k} x_\ell \right)_j \in L_p(\mathcal{M}, \varphi)[\ell_\infty].$$

(2)   If $p \geq 2$, then

$$\left(\frac{1}{j+1} \sum_{k=0}^{j} \frac{A_{j-k}^{r-1}}{A_j^r} \sum_{\ell \leq k} x_\ell \right)_j \in L_p(\mathcal{M}, \varphi)[\ell_\infty^c].$$

(3)   If $p \leq 2$, then the sequence from (1) decomposes into a sum

$$\frac{1}{j+1} \sum_{k} \frac{A_{j-k}^{r-1}}{A_j^r} \sum_{\ell \leq k} u_\ell + \frac{1}{j+1} \sum_{k} \frac{A_{j-k}^{r-1}}{A_j^r} \sum_{\ell \leq k} v_\ell,$$

where the first summand belongs to $L_p(\mathcal{M}, \varphi)[\ell_\infty^r]$ and the second one to $L_p(\mathcal{M}, \varphi)[\ell_\infty^c]$.

Almost sure and almost uniform convergence:

(4)   We have that

$$\lim_j \frac{1}{j+1} \sum_{k=0}^{j} \frac{A_{j-k}^{r-1}}{A_j^r} \sum_{\ell \leq k} x_\ell = 0$$

bilaterally φ-almost surely; if $p \geq 2$, then this convergence is φ-almost sure, and if $p \leq 2$, then it is row+column almost sure.

In particular, all these results hold in the case $r = 1$ which means ordinary Cesàro summation.

3.2.7   Coefficient Tests in the Algebra Itself

Assume again that $(\mathcal{M}, \varphi)$ is a fixed noncommutative probability space, i.e. $\mathcal{M}$ some von Neumann algebra together with a normal and faithful state φ with density D. We still intend to prove coefficient tests in the Haagerup L_p-space $L_p(\mathcal{M}, \varphi)$, but now for sequences of operators taken from the von Neumann algebra $\mathcal{M}$ itself. As in the semifinite case, we continue to consider φ-almost uniform and bilateral φ-almost uniform convergence. For sequences in the algebra itself φ-almost uniform convergence is well-known and widely used – may be first by Lance in [51] and [52] (for related studies see also [5, 19, 35, 37, 38, 71, 72, 73, 74, 75]).

The definition of row+column φ-almost uniform convergence seems to be new. For a sequence (y_n) in $\mathcal{M}$ and some y in $\mathcal{M}$ we say that:

- (y_n) converges to y φ-almost uniformly if for every $\varepsilon > 0$ there is a projection $s \in \mathcal{M}_{\mathrm{proj}}$ such that $\varphi(1 - s) < \varepsilon$ and

$$\lim_n \|(y_n - y)s\|_\infty = 0. \tag{3.63}$$

Moreover, we say that (y_n) is a φ-almost uniform Cauchy sequence whenever there is a projection s such that $\varphi(1 - s) < \varepsilon$ and

$$\lim_{n,m \to \infty} \|(y_n - y_m)s\|_\infty = 0. \tag{3.64}$$

- The sequence (y_n) converges to y bilaterally φ-almost uniformly if in (3.63) we only have

$$\lim_n \|s(y_n - y)s\|_\infty = 0,$$

and (y_n) is a bilateral φ-almost uniform Cauchy sequence if in (3.64) only

$$\lim_{n,m \to \infty} \|s(y_n - y_m)s\|_\infty = 0.$$

- (y_n) converges row+column φ-almost uniformly (is a row+column φ-almost uniform Cauchy sequence) provided the sequence $(y_n - y)$ decomposes into a sum $(a_n) + (b_n)$ of two sequences (a_n) and (b_n^*) where both converge φ-almost uniformly (both are φ-almost uniform Cauchy sequences).

As far as we know it is an open question whether or not every sequence (x_n) in $\mathcal{M}$ is φ-almost uniformly convergent to some $x \in \mathcal{M}$ provided the sequence $(x_n D^{1/2})$ from $L_2(\mathcal{M}, \varphi)$ converges φ-almost surely to $xD^{1/2}$; for partial solutions see [5, Corollary 2.1] and [38, p.100]. Nevertheless, the following proposition allows to deduce some of the coefficient tests we aim at, directly from the coefficient tests we got in the preceding section for Haagerup L_p's. The crucial tool which makes this possible is again a consequence of Goldstein's Lemma 24 and was first published in [42, Lemma 7.13]; for the sake of completeness we here copy the proof.

Proposition 9. *Let (x_n) be a sequence in $\mathcal{M}$.*

(1) If $(D^{1/2p}(x_n - x_m)D^{1/2p})_{nm} \in L_p(\mathcal{M})[c_0(\mathbb{N}_0^2)]$, then (x_n) is a bilateral φ-almost uniform Cauchy sequence.

(2) If $((x_n - x_m)D^{1/p})_{nm} \in L_p(\mathcal{M})[c_0^c(\mathbb{N}_0^2)]$, then (x_n) is a φ-almost uniform Cauchy sequence.

Proof. For (1) we as in the proof of Proposition 8 choose a, b, a_k, b_k and y_{nm} such that

$$D^{1/2p}(x_n - x_m)D^{1/2p} = ay_{nm}b \text{ and } \|a\|_{2p} < 1, \|b\|_{2p} < 1, \lim_{nm} \|y_{nm}\|_\infty = 0$$

and

$$a = \sum_k D^{1/2p} a_k \text{ and } \|D^{1/2p} a_k\|_{2p} < 2^{-k}$$

$$b = \sum_k b_k D^{1/2p} \text{ and } \|b_k D^{1/2p}\|_{2p} < 2^{-k}.$$

Next for each (n,m) choose some k_{nm} such that

$$\left\|D^{1/2p}(x_n - x_m)D^{1/2p} - \sum_{j,k=0}^{k_{nm}} D^{1/2p} a_j y_{nm} b_k D^{1/2p}\right\|_p \le 4^{-(n+m)},$$

and define

$$z_{nm} = (x_n - x_m) - \sum_{j,k=0}^{k_{nm}} a_j y_{nm} b_k.$$

Then by Hölder's inequality (see also (3.55))

$$\|D^{1/2} z_{nm} D^{1/2}\|_1 \le \|D^{1/2p} z_{nm} D^{1/2p}\|_p < 4^{-(n+m)}, \tag{3.65}$$

and if u_{nm} and v_{nm} denote the real and the imaginary part of z_{nm}, respectively, then (3.65) also holds for u_{nm} and v_{nm} instead of z_{nm}. Now we know from [23, Lemma1.2] (see also [26]) that there are positive elements u'_{nm} and u''_{nm} in $\mathcal{M}$ such that $u_{nm} = u'_{nm} + u''_{nm}$ and

$$\|D^{1/2} u_{nm} D^{1/2}\|_1 = \|D^{1/2} u'_{nm} D^{1/2}\|_1 + \|D^{1/2} u''_{nm} D^{1/2}\|_1 = \varphi(u'_{nm}) + \varphi(u''_{nm}).$$

Similarly, we obtain v'_{nm} and v''_{nm} for the imaginary part v_{nm} of z_{nm}, and hence from (3.65) (for u_{nm} and v_{nm}) that

$$\varphi(u'_{nm}) + \varphi(u''_{nm}) < 4^{-(n+m)} \text{ and } \varphi(v'_{nm}) + \varphi(v''_{nm}) < 4^{-(n+m)}.$$

Fix now $\varepsilon > 0$. Applying Goldstein's Lemma 24 to the family of all

$$a_n a_n^*, \ b_n^* b_n, \ u'_{nm}, \ u''_{nm}, \ v'_{nm}, \ v''_{nm}$$

we get a projection s in $\mathcal{M}$ such that $\varphi(1-s) < \varepsilon$ and for all n,m

$$\max\{\|s a_n a_n^* s\|_\infty, \|s b_n^* b_n s\|_\infty\} < 16\varepsilon^{-1} 2^{-n}$$

$$\max\{\|s u'_{nm} s\|_\infty, \|s u''_{nm} s\|_\infty, \|s v'_{nm} s\|_\infty, \|s v''_{nm} s\|_\infty\} < 16\varepsilon^{-1} 2^{-(n+m)}$$

(see also the proof of Proposition 8). But then we finally obtain

$$\|s(x_n - x_m)s\|_\infty \leq \|sz_{nm}s\|_\infty + \left\| \sum_{j,k=1}^{k_{nm}} s a_j y_{nm} b_k s \right\|_\infty$$

$$\leq \|s(u'_{nm} - u''_{nm})s\|_\infty + \|s(v'_{nm} - v''_{nm})s\|_\infty + \sum_{j,k=1}^{k_{nm}} \|s a_j y_{nm} b_k s\|_\infty$$

$$\leq 64\varepsilon^{-1} 2^{-(n+m)} + 16\varepsilon^{-1} \|y_{nm}\|_\infty \left(\sum_k 2^{-k/2} \right)^2 \to 0 \text{ as } n, m \to \infty,$$

the conclusion in (1). The proof of (2) is similar.   □

The following theorem is the main result in this section – compare with Theorem 20 (tracial case) and Theorem 26 (nontracial case). This time the corresponding maximal inequalities will be given later in Sect. 3.2.9.

Theorem 30. *Assume that S is a summation method and ω a Weyl sequence such that for each orthonormal series $\sum_k \alpha_k x_k$ in $L_2(\mu)$ we have*

$$\sup_j \left| \sum_{k=0}^\infty s_{jk} \sum_{\ell=0}^k \frac{\alpha_\ell}{\omega_\ell} x_\ell \right| \in L_2(\mu) .$$

Then for any sequence (x_k) in $\mathcal{M}$ and $1 \leq p < \infty$ the following holds:

(1) If the series $\sum_k D^{1/2p} x_k D^{1/2p}$ is unconditionally convergent in $L_p(\mathcal{M}, \varphi)$, then

$$\left(\sum_{k=0}^\infty s_{jk} \sum_{\ell=0}^k \frac{x_\ell}{\omega_\ell} \right)_j \tag{3.66}$$

is a bilateral φ-almost uniform Cauchy sequence.

(2) If $p \geq 2$ and the series $\sum_k x_k D^{1/p}$ converges unconditionally in $L_p(\mathcal{M}, \varphi)$, then the linear means from (3.66) form a φ-almost uniform Cauchy sequence.

(3) If $p \leq 2$ and the series $\sum_k D^{1/2p} x_k D^{1/2p}$ converges unconditionally in $L_p(\mathcal{M}, \varphi)$, then the linear means from (3.66) form a row+column φ-almost uniform Cauchy sequence.

Proof. Use first Theorem 26 to show as in the proof of Lemma 27 that (3.57) holds. Clearly, the conclusion then follows from Proposition 9.   □

As in the preceding sections we apply our results to ordinary, Riesz, Cesàro, and Abel summation of unconditionally convergent series – recall Corollary 6 (commutative case), Corollary 10 (tracial case) and Corollary 14 (nontracial case) which are all slightly different.

Corollary 17. *Let* (x_k) *be a sequence in* $\mathcal{M}$ *such that the series* $\sum_k D^{1/2p} x_k D^{1/2p}$ *is unconditionally convergent in* $L_p(\mathcal{M}, \varphi)$. *Then the following results hold:*

(1) $\left(\sum_{k=0}^{j} \dfrac{x_k}{\log k} \right)_j$ *is a bilateral* φ-*almost uniform Cauchy sequence*

(2) $\left(\sum_{k=0}^{j} \dfrac{\lambda_{k+1} - \lambda_k}{\lambda_{j+1}} \sum_{\ell \le k} \dfrac{x_\ell}{\log\log \lambda_\ell} \right)_j$ *is a bilateral* φ-*almost uniform Cauchy sequence for every strictly increasing, unbounded and positive sequence* (λ_k) *of scalars*

(3) $\left(\sum_{k=0}^{j} \dfrac{A_{j-k}^{r-1}}{A_j^r} \sum_{\ell \le k} \dfrac{x_\ell}{\log\log \ell} \right)_j$ *is a bilateral* φ-*almost uniform Cauchy sequence for every* $r > 0$

(4) $\left(\sum_{k=0}^{\infty} \rho_j^k \dfrac{x_k}{\log\log k} \right)_j$ *is a bilateral* φ-*almost uniform Cauchy sequence for every positive strictly increasing sequence* (ρ_j) *converging to* 1

(5) If $p \ge 2$ *and* $\sum_k x_k D^{1/p}$ *is unconditionally convergent in* $L_p(\mathcal{M}, \varphi)$, *then the series in* (1)–(4) *are* φ-*almost uniform Cauchy sequences.*

(6) If $p \le 2$ *and* $\sum_k D^{1/2p} x_k D^{1/2p}$ *is unconditionally convergent in* $L_p(\mathcal{M}, \varphi)$, *then the series in* (1)–(4) *are row+column* φ-*almost uniform Cauchy sequences.*

Again the logarithmic terms can be omitted under restrictions on the series; compare with Theorem 27 (nontracial case), Theorem 21 (tracial case) and Theorem 14 (commutative case).

Theorem 31. *Let S be a summation method. Then for* $1 \le q < p < \infty$, *each sequence* (x_k) *in* $\mathcal{M}$ *and each* $\alpha \in \ell_q$ *the following statements hold true:*

(1) If $(D^{1/2p} x_k D^{1/2p})$ *is a weakly* q'-*summable sequence in* $L_p(\mathcal{M}, \varphi)$, *then*

$$\left(\sum_{k=0}^{\infty} s_{jk} \sum_{\ell \le k} \alpha_\ell x_\ell \right)_j \tag{3.67}$$

is a bilateral φ-*almost uniform Cauchy sequence.*

(2) If $2 \le p$ *and* $(x_k D^{1/p})$ *is a weakly* q'-*summable sequence in* $L_p(\mathcal{M}, \varphi)$, *then the sequence from* (3.67) *even forms a* φ-*almost uniform Cauchy sequence.*

Proof. As before, we first use statement (1) of Theorem 27 to show (3.57) (as explained in the proof of Lemma 27). The conclusion then follows from Proposition 9.   □

Like in Corollary 7 (commutative case) and Corollary 11 (tracial case) this result can be applied to ordinary, Cesàro, Riesz, and Abel summation.

From Theorem 30, Corollary 17 or Theorem 31 we can only conclude that the linear means $(u_j) = \left(\sum_{k=0}^{\infty} s_{jk} \sum_{\ell=0}^{k} \dfrac{x_\ell}{\omega_\ell} \right)_j$ are φ-almost uniform Cauchy sequences and not φ-almost uniform convergent sequences. Why? Going back to the commutative case $\mathcal{M} = L_\infty(\mu)$ we immediately understand that the linear means in general will not converge μ-almost everywhere to some element from $L_\infty(\mu)$ itself.

Conjecture 1. Let (u_n) be a sequence in $\mathcal{M}$. Then for each $1 \le p < \infty$ the following statements hold:

(1) If (u_n) is a bilateral φ-almost uniform Cauchy sequence in $\mathcal{M}$ which in $L_p(\mathcal{M}, \varphi)$ converges to some $D^{1/2p} u D^{1/2p}$ with $u \in \mathcal{M}$, then (u_n) converges to u bilaterally φ-almost uniformly.
(2) If (u_n) is a φ-almost uniform Cauchy sequence in $\mathcal{M}$ which in $L_p(\mathcal{M}, \varphi)$ converges to some $u D^{1/p}$ with $u \in \mathcal{M}$, then (u_n) converges to u φ-almost uniformly.

We include a short discussion which shows that for certain sequences (u_n) (or sometimes even for all such sequences) this conjecture holds, and then the preceding results can be improved (provided the linear means u_j from $\mathcal{M}$ converges in $L_p(\mathcal{M}, \varphi)$ to some element from $\mathcal{M}$). Given $1 \le p < \infty$, we say that a sequence (u_n) from $\mathcal{M}$ satisfies condition (D_p^{col}) if $\|u_n D^{1/p}\|_p \to 0$ and for any projection q the following implication holds true:

$$(u_n q) \text{ converges in } \mathcal{M} \;\Rightarrow\; u_n q \to 0 \text{ in } \mathcal{M};$$

similarly, (u_n) satisfies (D_p^{sym}) whenever $\|D^{1/2p} u_n D^{1/2p}\|_p \to 0$ and for any q

$$(q u_n q) \text{ converges in } \mathcal{M} \;\Rightarrow\; q u_n q \to 0 \text{ in } \mathcal{M}.$$

Obviously, if in Conjecture 1, (1) we add the assumption (D_p^{sym}) and in (2) the assumption (D_p^{col}), then both implications hold true.

For example, every sequence (u_n) of selfadjoint operators satisfies (D_p^{col}); but unfortunately not every sequence (u_n) satisfies (D_2^{col}) (as pointed out to me by S. Goldstein): Take some $\xi \in \ell_2$ such that $|\xi_n|^{-2} 2^{-n} \to 0$ and $\|\xi\|_2 = 1$. Consider on the von Neumann algebra $\mathcal{M} = \mathcal{L}(\ell_2)$ the normal faithful state $\varphi = \sum_k \frac{1}{2^k} e_k \otimes e_k$. Define the operators $u_n = \frac{1}{\xi_n} e_n \otimes e_1$ and the orthogonal projection $q = \xi \otimes \xi$. Then it can be seen easily that $\varphi(u_n^* u_n) = 1/(|\xi_n|^2 2^n) \to 0$, hence $u_n D^{1/2} \to 0$ in $L_2(\mathcal{M}, \varphi)$. But we have that $u_n q = \xi \otimes e_1$, i.e. $u_n q \to \xi \otimes e_1$ in $\mathcal{M}$, although $\xi \otimes e_1 \ne 0$.

On the other hand, it can be seen easily that in von Neumann algebras $\mathcal{M}$ in which all projections q are analytic every sequence (u_n) satisfies (D_p^{sym}) and (D_p^{col}) (recall from [89] that an element $u \in \mathcal{M}$ is analytic if the function $t \in \mathbb{R} \to \sigma_t^{\varphi}(u)$ $\in \mathcal{M}$ extends to an analytic function on $\mathbb{C}$); indeed, let (u_n) be a sequence in $\mathcal{M}$ such that $\|D^{1/2p} u_n D^{1/2p}\|_p \to 0$ and $q u_n q \to y$ for some $y \in \mathcal{M}$ and some projection q. Then we deduce from [48, Remark 6.2] that

$$D^{1/2p}(q u_n q - y) D^{1/2p} = \sigma_{-i/2p}^{\varphi}(q) D^{1/2p} u_n D^{1/2p} \sigma_{-i/2p}^{\varphi}(q)^* - D^{1/2p} y D^{1/2p}.$$

But now the left side converges to 0 and the right side converges to $-D^{1/2p} y D^{1/2p}$. This implies that $D^{1/2p} y D^{1/2p} = 0$, and therefore $y = 0$. In the column case we argue in the same way. See [89] for examples in which all projections are analytic.

Of course, the preceding discussion can be applied to Theorem 30 and leads in certain situations to an improvement, e.g. in the column case for a selfadjoint sequence $\left(\sum_{k=0}^{\infty} s_{jk} \sum_{\ell=0}^{k} x_{\ell} / \omega_{\ell} \right)_j$ of linear means. We illustrate this again looking at the Menchoff-Rademacher theorem from (1.1) which is still the prototype of the results we are interested in. Here is a noncommutative variant entirely formulated in terms of the given algebra and state itself; recall from (1.15) that on $\mathcal{M}$ the scalar product $(x|y) := \varphi(y^*x)$ leads to the prehilbert space $(\mathcal{M}, \|\cdot\|_2)$ (the GNS-construction).

Corollary 18. *Assume that* (x_k) *is a* φ-*orthogonal sequence in* $\mathcal{M}$ *such that*

$$\sum_k \log^2 k \, \|x_k\|_2^2 < \infty.$$

If the sum x of the $\|\cdot\|_2$-*convergent series* $\sum_k x_k$ *belongs to* $\mathcal{M}$ *and this series satisfies* (D_2^{col}) *(e.g. if all x_k are selfadjoint), then*

$$x = \sum_{k=0}^{\infty} x_k \quad \varphi\text{-almost uniformly.}$$

Proof. We do the proof in the Haagerup Hilbert space $L_2(\mathcal{M}, \varphi)$ which is the completion of $(\mathcal{M}, \|\cdot\|_2)$. By assumption we know that the series

$$\sum_k \log k \, \|x_k D^{1/2}\|_2 \, \frac{x_k D^{1/2}}{\|x_k D^{1/2}\|_2}$$

is orthonormal and hence unconditionally convergent in $L_2(\mathcal{M}, \varphi)$. But then we deduce from Corollary 17, (5) that the sequence $\left(\sum_{k=0}^{j} x_k D^{1/2} \right)_j$ is a φ-almost uniform Cauchy sequence. Since, again by assumption, we in $L_2(\mathcal{M}, \varphi)$ have that $\sum_{k=0}^{\infty} x_k D^{1/2} = x D^{1/2}$, the proof completes (see the above discussion). $\square$

Of course, similar arguments applied to Corollary 17 and Theorem 31 give other interesting noncommutative coefficient tests. In particular, all six statements of Corollary 17 on Cesàro, Riesz, and Abel summation lead to natural extensions of classical results which can be entirely formulated in terms of the underlying noncommutative probability space $(\mathcal{M}, \varphi)$; for example to another noncommutative extension of the Kaczmarz-Menchoff theorem.

Corollary 19. *Assume that* (x_k) *is a* φ-*orthogonal sequence in* $\mathcal{M}$ *such that*

$$\sum_k (\log \log k)^2 \, \|x_k\|_2^2 < \infty.$$

If the sum x of the $\|\cdot\|_2$-*convergent series* $\sum_k x_k$ *belongs to* $\mathcal{M}$ *and this series satisfies* (D_2^{col}) *(e.g. if all x_k are selfadjoint), then*

$$\sum_k x_k = \frac{1}{j+1} \sum_{k=0}^{j} \sum_{\ell=0}^{k} x_{\ell} \quad \varphi\text{-almost uniformly.} \tag{3.68}$$

3.2.8 Laws of Large Numbers in the Algebra Itself

We now study for arbitrary summation methods S and for sequences (x_k) in the algebra $\mathcal{M}$ itself laws of large numbers

$$\lim_j \frac{1}{j+1} \sum_{k=0}^{j} s_{jk} \sum_{\ell=0}^{k} x_\ell = 0 \quad \varphi\text{-almost uniformly};$$

compare with the results from the Sects. 2.3.2 (commutative case), 3.1.7 (tracial case), and 3.2.6 (nontracial case). As above we fix some von Neumann algebra $\mathcal{M}$ of operators acting on the Hilbert spaces H, and a faithful and normal state φ on $\mathcal{M}$ with density D.

For φ-orthogonal sequences in $\mathcal{M}$ the following law of large numbers of Jajte from [36, 37, Sect. 4.4.1] is an analog of (3.62) and its strong extension from Theorem 28: *Let (x_k) be a sequence in $\mathcal{M}$ which is φ-orthogonal and satisfies*

$$\sum_k \frac{\log^2 k}{k^2} \varphi(|x_k|^2) < \infty.$$

Then

$$\lim_j \frac{1}{j+1} \sum_{k=0}^{j} x_k = 0 \quad \varphi\text{-almost uniformly.} \tag{3.69}$$

Unfortunately, we cannot fully recover this result within our setting (since we do not know wether Conjecture 1 holds without any further assumption on (u_n) like (D_p^{col}) or (D_p^{sym})). Nevertheless, the following theorem can be considered as a far reaching complement of the above described work of Jajte.

Theorem 32. *Let S be a lower triangular summation method. Assume that ω is an increasing sequence of positive scalars such that for each orthogonal sequence (x_k) in $L_2(\mu)$ with $\sum_k \frac{\omega_k^2}{k^2}\|x_k\|_2^2 < \infty$ we have*

$$\sup_j \left| \frac{1}{j+1} \sum_{k=0}^{j} s_{jk} \sum_{\ell=0}^{k} x_\ell \right| \in L_2(\mu).$$

Then for every sequence (x_k) in $\mathcal{M}$ for which the series $\sum_k \frac{\omega_k}{k} D^{1/2p} x_k D^{1/2p}$ is unconditionally convergent in $L_p(\mathcal{M}, \varphi)$ we have that

(1) $\lim_j \frac{1}{j+1} \sum_{k=0}^{j} s_{jk} \sum_{\ell=0}^{k} D^{1/2p} x_\ell D^{1/2p} = 0$ *in* $L_p(\mathcal{M}, \varphi)$

(2) $\left(\frac{1}{j+1} \sum_{k=0}^{j} s_{jk} \sum_{\ell \le k} x_\ell \right)_j$ *is a bilateral φ-almost uniform Cauchy sequence, and if this sequence satisfies condition (D_p^{sym}), then it converges to zero bilaterally φ-almost uniformly.*

In the column case a similar result holds – if $p \geq 2$, then the linear means from (2) form a φ-almost uniform Cauchy sequence which under the condition of (D_p^{col}) is even a φ-almost uniform zero sequence.

Proof. We know from Theorem 28 that for every unconditionally convergent series $\sum_k y_k = \sum_k \frac{\omega_k}{k} \frac{k}{\omega_k} y_k$ in $L_p(\mathcal{M})$ we have

$$\left(\sum_{k=0}^{j} \left(\frac{k}{(j+1)\omega_k} \sum_{\ell=k}^{j} s_{j\ell} \right) y_k \right)_j = \left(\frac{1}{j+1} \sum_{k=0}^{j} s_{jk} \sum_{\ell \leq k} \frac{\ell}{\omega_\ell} y_\ell \right)_j \in L_p(\mathcal{M})[\ell_\infty]$$

(compare with the proof of Theorem 22). Hence we deduce as in the proof of Lemma 27 (or better, the fact that the assumption in Lemma 27 implies (3.57)) that

$$\left(\frac{1}{i+1} \sum_{k=0}^{i} s_{ik} \sum_{\ell \leq k} \frac{\ell}{\omega_\ell} y_\ell - \frac{1}{j+1} \sum_{k=0}^{j} s_{jk} \sum_{\ell \leq k} \frac{\ell}{\omega_\ell} y_\ell \right)_{i,j} \in L_p(\mathcal{M})[c_0(\mathbb{N}_0^2)].$$

Therefore by Proposition 9 and the unconditionality of the series $\sum_k \frac{\omega_k}{k} D^{1/2p} x_k D^{1/2p}$ we know that the sequence (u_j) in $\mathcal{M}$ given by

$$u_j = \frac{1}{j+1} \sum_{k=0}^{j} s_{jk} \sum_{\ell \leq k} x_\ell$$

is a bilaterally φ-almost uniform Cauchy sequence. On the other hand, since S is a summation method and the series $\sum_k \frac{D^{1/2p} x_k D^{1/2p}}{k}$ converges in $L_p(\mathcal{M})$,

$$\sum_{k=0}^{\infty} \frac{D^{1/2p} x_k D^{1/2p}}{k} = \lim_j \sum_{k=0}^{j} s_{jk} \sum_{\ell=0}^{k} \frac{D^{1/2p} x_\ell D^{1/2p}}{\ell},$$

the limit taken in $L_p(\mathcal{M})$. Then by Kronecker's Lemma 11, (2) we see that in $L_p(\mathcal{M})$

$$0 = \lim_j \frac{1}{j+1} \sum_{k=0}^{j} s_{jk} \sum_{\ell=0}^{k} D^{1/2p} x_\ell D^{1/2p} = \lim_j D^{1/2p} u_j D^{1/2p}$$

(compare with the proof of Theorem 15). Hence (1) and the first statement in (2) are proved – the second statement in (2) then follows from the discussion given after Conjecture 1. Similar arguments lead to the conclusion on the column case. □

Of course, Theorem 32 again can be applied to ordinary, Riesz, Cesàro, and Abel summation. If we combine it with Theorem 16 (the case of vector-valued Banach function spaces), then we get an analog of Theorem 23 (tracial case) and Theorem 29 (nontracial case) – again no log-terms are involved.

Corollary 20. *Let* (x_k) *be a sequence in* $\mathcal{M}$ *such that* $\sum_k \frac{1}{k} D^{1/2p} x_k D^{1/2p}$ *converges unconditionally in* $L_p(\mathcal{M}, \varphi)$ *and let* $r > 0$. *Then the sequence*

$$\left(\frac{1}{j+1} \sum_{k=0}^{j} \frac{A_{j-k}^{r-1}}{A_j^r} \sum_{\ell \leq k} x_\ell \right)_j$$

of linear means is a bilateral φ-almost uniform Cauchy sequence which converges to zero bilaterally φ-almost uniformly whenever it satisfies condition (D_p^{sym}). *In the column case a similar result holds – if $p \geq 2$, then the linear means even form a φ-almost uniform Cauchy sequence which under the condition* (D_p^{col}) *is a φ-almost uniform zero sequence.*

3.2.9 Maximal Inequalities in the Algebra Itself

Finally we show Kantorovich-Menchoff-Rademacher type inequalities for elements of the von Neumann algebra itself – recall that in the preceding section we obtained coefficient tests and laws of large numbers for sequences of operators in $\mathcal{M}$ without proving any corresponding maximal inequalities. Let us again fix a von Neumann algebra $\mathcal{M}$ and a faithful normal state φ with density D.

To give an idea of what we plan to do, we state the following Kantorovich-Menchoff-Rademacher type inequality for φ-orthonormal operators (already announced in the introduction in (1.17)) – this inequality is the natural counterpart of the Corollaries 17, (1) and 18. Later this result will be proved as an immediate consequence of the much more general Theorem 33; for the classical Kantorovich-Menchoff-Rademacher inequality see (2.28).

Corollary 21. *For each choice of finitely many φ-orthonormal operators* $x_0, \dots, x_n \in \mathcal{M}$ *and each choice of finitely many scalars* $\alpha_0, \dots \alpha_n$ *there is a factorization*

$$\sum_{k=0}^{j} \alpha_k x_k = z_j c, \quad 0 \leq j \leq n$$

with $z_j, c \in \mathcal{M}$ *such that*

$$\sup_j \| z_j \|_\infty \, \varphi(c^* c)^{1/2} \leq C \left(\sum_{k=0}^{n} |\alpha_k \log k|^2 \right)^{1/2},$$

$C > 0$ *some constant.*

The proof of this result is given at the end of this section; the strategy is to deduce it from its counterparts in Haagerup L_p's. First, we need a variant of the maximal norms already defined in symmetric spaces of measurable operators (see Sect. 3.1.2) and Haagerup L_p-spaces (see Sect. 3.2.2). For a finite sequence (x_i) in $\mathcal{M}$ (a sequence which is zero from some index on) and $1 \leq p \leq \infty$ put

$$\||(x_i)\||_p^{\mathrm{sym}} := \inf \|D^{1/2p}c\|_{2p} \sup_i \|z_i\|_\infty \|dD^{1/2p}\|_{2p},$$

where the infimum is taken over all uniform factorizations

$$x_i = c z_i d, \quad i \in \mathbb{N}$$

such that $c, d \in \mathcal{M}$ and (z_i) is a finite sequence in $\mathcal{M}$. If we just consider factorizations $x_i = z_i d$ or $x_i = c z_i$, then we write

$$\|| \cdot \||_p^{\mathrm{col}} \quad \text{or} \quad \|| \cdot \||_p^{\mathrm{row}},$$

respectively. Moreover, if we only consider the preceding three norms on $\prod_{k=0}^n \mathcal{M}$, then we will write

$$\|| \cdot \||_{p,n}^{\mathrm{sym}}, \|| \cdot \||_{p,n}^{\mathrm{col}}, \text{ and } \|| \cdot \||_{p,n}^{\mathrm{row}},$$

respectively. The following result is the counterpart of the Propositions 3 and 7.

Proposition 10. $\|| \cdot \||_{p,n}^{\mathrm{sym}}$ *is a norm on* $\prod_{k=0}^n \mathcal{M}$, *and in the column and row case this is true provided* $p \geq 2$.

For $\alpha = \mathrm{sym}$, col or row we write

$$\mathcal{M} \otimes_p^\alpha \ell_\infty^n := \left(\prod_{k=0}^n \mathcal{M}, \|| \cdot \||_{p,n}^\alpha \right).$$

Proof. Of course, the proof of this result is similar to that of Proposition 3 – for the sake of completeness we check that $\|| \cdot \||_{p,n}^{\mathrm{sym}}$ satisfies the triangle inequality. Take two finite sequences $(x_k(i))_{1 \leq i \leq n}$, $k = 1, 2$ in $\mathcal{M}$ which allow uniform factorizations

$$x_k(i) = c_k z_k(i) d_k$$

$$c_k, z_k(i), d_k \in \mathcal{M}$$

$$\|D^{1/2p}c_k\|_{2p}^2 = \|d_k D^{1/2p}\|_{2p}^2 = \||x_k\||_{p,n}^{\mathrm{sym}}$$

$$\sup_i \|z_k(i)\|_\infty \leq 1.$$

By taking polar decompositions we may assume that the c_k's and d_k's are positive. Define

$$\tilde{c} := \left(D^{1/2p}(c_1^2 + c_2^2) D^{1/2p} \right)^{1/2} \in L_{2p}(\mathcal{M}),$$

$$\tilde{d} := \left(D^{1/2p}(d_1^2 + d_2^2) D^{1/2p} \right)^{1/2} \in L_{2p}(\mathcal{M});$$

clearly,

$$\|\tilde{d}\|_{2p} = \|D^{1/2p}d_1^2 D^{1/2p} + D^{1/2p}d_2^2 D^{1/2p}\|_p^{1/2}$$

$$\leq \left(\|D^{1/2p}d_1^2 D^{1/2p}\|_p + \|D^{1/2p}d_2^2 D^{1/2p}\|_p\right)^{1/2}$$

$$= \left(\|(D^{1/2p}d_1)^2\|_p + \|(D^{1/2p}d_2)^2\|_p\right)^{1/2}$$

$$= \left(\|D^{1/2p}d_1\|_{2p}^2 + \|D^{1/2p}d_2\|_{2p}^2\right)^{1/2} = \left(\|\|x_1\|\|_{p,n}^{sym} + \|\|x_2\|\|_{p,n}^{sym}\right)^{1/2},$$

and similarly

$$\|\tilde{c}\|_{2p} \leq \left(\|\|x_1\|\|_{p,n}^{sym} + \|\|x_2\|\|_{p,n}^{sym}\right)^{1/2}.$$

Since φ is faithful, the range projection of D and also $D^{1/2p}$ equals 1. Moreover,

$$\tilde{c}^2 \leq \|c_1^2 + c_2^2\|_\infty D^{1/p}$$

$$\tilde{d}^2 \leq \|d_1^2 + d_2^2\|_\infty D^{1/p};$$

indeed, in $L_0(\mathscr{R}, \tau)$ we have

$$(\tilde{d}^2\xi \mid \xi) = \left((d_1^2 + d_2^2)D^{1/2p}\xi \mid D^{1/2p}\xi\right)$$

$$= \|(d_1^2 + d_2^2)^{1/2}D^{1/2p}\xi\|^2 \leq \|d_1^2 + d_2^2\|_\infty (D^{1/p}\xi \mid \xi).$$

Define

$$c := D^{-1/2p}\tilde{c}, \quad d := \tilde{d}D^{-1/2p} \in \mathscr{M}$$

(see Proposition 7). Moreover, it follows from

$$c^2 = D^{-1/2p}\tilde{c}^2 D^{-1/2p} = c_1^2 + c_2^2$$

$$d^2 = D^{-1/2p}\tilde{d}^2 D^{-1/2p} = d_1^2 + d_2^2$$

that there are contractions $a_k, b_k \in \mathscr{M}$ such that

$$c_k = c a_k \quad \text{and} \quad d_k = b_k d$$

$$a_1 a_1^* + a_2 a_2^* = r(c^2) \quad \text{and} \quad b_1^* b_1 + b_2^* b_2 = r(d^2).$$

Finally, we define $z(\cdot) = a_1 z_1(\cdot)b_1 + a_2 z_2(\cdot)b_2$, and conclude (exactly as in the proof of Proposition 3) that

$$x_1 + x_2 = c a_1 z_1(\cdot)b_1 d + c a_2 z_2(\cdot)b_2 d = c z(\cdot)d$$

and

$$\||x_1 + x_2\||_{p,n}^{\text{sym}} = \||cz(\cdot)d\||_{p,n}^{\text{sym}}$$

$$\le \|D^{1/2p}c\|_{2p} \sup_i \|z(i)\|_\infty \|dD^{1/2p}\|_{2p}$$

$$= \|\tilde{c}\|_{2p} \sup_i \|z(i)\|_\infty \|\tilde{d}\|_{2p} \le \||x_1\||_{p,n}^{\text{sym}} + \||x_2\||_{p,n}^{\text{sym}},$$

the conclusion. The column and the row case follow from similar arguments.   □

The next lemma will help to understand that the completion of $\mathscr{M} \otimes_p^\alpha \ell_\infty^n$.

Lemma 28. *For $x \in \mathscr{M}$ we have*

$$\||x\||_{p,1}^{\text{sym}} = \inf_{x=czd,\, c,z,d\in\mathscr{M}} \|D^{1/2p}c\|_{2p}\|z\|_\infty\|dD^{1/2p}\|_{2p} = \|D^{1/2p}xD^{1/2p}\|_p,$$

and for $p \ge 2$ a similar result holds in the row and column case.

Proof. This time we prefer to do the column case first since it is much easier than the symmetric one. Clearly,

$$\||x\||_{p,1}^{\text{col}} \le \|1\|_\infty \|xD^{1/p}\|_p = \|xD^{1/p}\|_p,$$

and conversely for each decomposition $x = zd$ with $z,d \in \mathscr{M}$ we have

$$\|xD^{1/p}\|_p = \|zdD^{1/p}\|_p \le \|z\|_\infty\|dD^{1/p}\|_p.$$

For the symmetric case take $x' \in \left(\mathscr{M}, \||\cdot\||_{p,1}^{\text{sym}}\right)'$ with $\|x'\| \le 1$ and $x'(x) = \||x\||_{p,1}^{\text{sym}}$. Define the contraction

$$\psi : L_{2p}(\mathscr{M})\tilde{\otimes}_\pi L_{2p}(\mathscr{M}) \longrightarrow \mathbb{C},$$

first on $D^{1/2p}\mathscr{M} \otimes \mathscr{M}D^{1/2p}$ by

$$\psi(D^{1/2p}a \otimes bD^{1/2p}) = x'(ab),$$

and then by continuous extension on the whole completion $L_{2p}(\mathscr{M})\tilde{\otimes}_\pi L_{2p}(\mathscr{M})$ of the projective tensor product. Clearly, we have

$$\psi(ax \otimes b) = \psi(a \otimes xb) \quad \text{for all } a,b \in L_{2p}(\mathscr{M}) \text{ and } x \in \mathscr{M}.$$

But then the mapping

$$T : L_{2p}(\mathcal{M}) \to L_{(2p)'}(\mathcal{M}) = L_{2p}(\mathcal{M})'$$

given by

$$T(a)(b) = \psi(a \otimes b)$$

is a well-defined linear operator which has norm ≤ 1 and satisfies

$$T(ax) = T(a)x \text{ for all } a \in L_{2p}(\mathcal{M}) \text{ and } x \in \mathcal{M}.$$

Hence, by [40, Theorem 2.5] there is $v \in L_{p'}(\mathcal{M})$ with $\|v\|_{p'} \leq 1$ such that

$$T(a) = va \quad \text{for all } a \in L_{2p}(\mathcal{M}),$$

which gives

$$\psi(a \otimes b) = T(a)(b) = \text{tr}(vab) \quad \text{for all } a, b \in L_{2p}(\mathcal{M}).$$

This finally by Hölder's inequality implies that

$$\|x\|_{p,1}^{\text{sym}} = x'(x) = |\psi(D^{1/2p}x \otimes 1D^{1/2p})|$$

$$= |\text{tr}(vD^{1/2p}xD^{1/2p})| \leq \|D^{1/2p}xD^{1/2p}\|_p,$$

the conclusion.   □

By definition and the triangle inequality we have that for $\alpha = \text{sym}, \text{row}$ or col and every choice of operators $x_0, \ldots, x_n \in L_p(\mathcal{M})$

$$\max_{0 \leq k \leq n} \|x_k\|_{p,1}^\alpha \leq \|(x_k)\|_{p,n}^\alpha \leq \sum_{k=0}^{n} \|x_k\|_{p,1}^\alpha.$$

Hence for the completions we conclude from Lemma 23 and Lemma 28 the following algebraic equalities:

$$\mathcal{M} \tilde{\otimes}_p^\alpha \ell_\infty^n := \text{completion of } \mathcal{M} \otimes_p^\alpha \ell_\infty^n$$

$$= \prod_{k=0}^{n} \text{completion of } \left(\mathcal{M}, \| \cdot \|_{p,1}^\alpha\right) = \prod_{k=0}^{n} L_p(\mathcal{M}).$$

Clearly, this gives that the canonical contraction

$$i : \mathcal{M} \otimes_p^{\text{sym}} \ell_\infty^n \longrightarrow L_p(\mathcal{M})[\ell_\infty^n], \ i(x_k) := (D^{1/2p}x_k D^{1/2p})$$

extends to a bijective contraction j from $\mathcal{M} \tilde{\otimes}_p^{\text{sym}} \ell_\infty^n$ to $L_p(\mathcal{M})[\ell_\infty^n]$, and similarly in the row and column case. More can be said:

Proposition 11. *The canonical mapping j is a metric bijection, i.e. the equality*

$$\mathcal{M} \tilde{\otimes}_p^{sym} \ell_\infty^n = L_p(\mathcal{M}, \varphi)[\ell_\infty^n]$$

holds isometrically; a similar result holds in the row and column case provided $p \geq 2$.

Proof. We check – in the symmetric case only – that j is a metric bijection. Take $y_0, ..., y_n \in L_p(\mathcal{M})[\ell_\infty^n]$ with $\|(y_k)\|_{L_p[\ell_\infty^n]} < 1$ and choose a factorization

$$y_k = c z_k d, 0 \leq k \leq n, \quad \text{and} \quad \|c\|_{2p} < 1, \ \|d\|_{2p} < 1, \ \|z_k\|_\infty < 1.$$

Approximate

$$c = \sum_{l=0}^{\infty} D^{1/2p} c_l \quad \text{with} \quad \|D^{1/2p} c_l\|_{2p} \leq \frac{1}{2^l},$$

$$d = \sum_{l=0}^{\infty} d_l D^{1/2p} \quad \text{with} \quad \|d_l D^{1/2p}\|_{2p} \leq \frac{1}{2^l},$$

and put for m, k

$$x_k^m = \left(\sum_{l=0}^{m} c_l \right) z_k \left(\sum_{l=0}^{m} d_l \right) \in \mathcal{M}.$$

Then $((x_k^m)_{k=0}^n)_m$ is a Cauchy sequence in $\mathcal{M} \otimes_p^{sym} \ell_\infty^n$ – indeed

$$\left\| (x_k^m)_k - (x_k^{m+1})_k \right\|_{p,n}^{sym} = \left\| \left(\left(\sum_{l \leq m} c_l \right) z_k \left(\sum_{l \leq m} d_l \right) - \left(\sum_{l \leq m+1} c_l \right) z_k \left(\sum_{l \leq m+1} d_l \right) \right)_k \right\|_{p,n}^{sym}$$

$$\leq \left\| (c_{m+1} z_k d_{m+1}) \right\|_{p,n}^{sym} + \left\| \left(c_{m+1} z_k \sum_{l \leq m} d_l \right) \right\|_{p,n}^{sym} + \left\| \left(\sum_{l \leq m} c_l z_k d_{m+1} \right) \right\|_{p,n}^{sym}$$

$$\leq \frac{5}{2^{m+1}}.$$

Put $y := \lim_m (x_k^m)_k \in \mathcal{M} \tilde{\otimes}_p^{sym} \ell_\infty^n$. Then clearly

$$j(y) = \lim_m j(x_k^m)_k = \lim_m \left(\sum_{l=0}^{m} D^{1/2p} c_l z_k \sum_{l=0}^{m} d_l D^{1/2p} \right)_k$$

$$= \left(\lim_m \left(\sum_{l=0}^{m} D^{1/2p} c_l z_k \sum_{l=0}^{m} d_l D^{1/2p} \right) \right)_k$$

$$= \left(\sum_{l=0}^{\infty} D^{1/2p} c_l z_k \sum_{l=0}^{\infty} d_l D^{1/2p} \right)_k = \left(c z_k d \right)_k = (y_k)_k.$$

Moreover, we have for each m

$$\||y\||_{p,n}^{\text{sym}} \leq \||y - (x_k^m)\||_{p,n}^{\text{sym}} + \||(x_k^m)\||_{p,n}^{\text{sym}}$$

$$\leq \||y - (x_k^m)\||_{p,n}^{\text{sym}} + \Big\| \sum_{l=0}^{m} D^{1/2p} c_l \Big\|_{2p} \sup_k \|z_k\|_\infty \Big\| \sum_{l=0}^{m} d_l D^{1/2p} \Big\|_{2p},$$

hence for $m \to \infty$ we obtain as desired

$$\|y\|_{p,n}^{\text{sym}} \leq \Big\| \sum_{l=0}^{m} D^{1/2p} c_l \Big\|_{2p} \sup_k \|z_k\|_\infty \Big\| \sum_{l=0}^{m} d_l D^{1/2p} \Big\|_{2p} < 1,$$

and the proof is complete.   □

We now are in a position to state and to prove the two main statements of this section. The first one is the following analog of Theorem 13 (commutative case), 17 (tracial case), and Theorem 24 (nontracial case).

Theorem 33. *Let $A = (a_{jk})$ be a $(2,2)$-maximizing matrix. Then there is a constant $C > 0$ such for each n and each choice of finitely many $x_0, \ldots, x_m \in \mathcal{M}$ we have*

$$\Big\| \Big(\sum_{k=0}^{m} a_{jk} x_k \Big)_{0 \leq j \leq n} \Big\|_{p,n}^{\text{sym}} \leq C \sup_{|\varepsilon_k|=1} \Big\| \sum_{k=0}^{m} \varepsilon_k x_k \Big\|_p,$$

i.e. for each choice of finitely many linear means $(\sum_{k=0}^{m} a_{jk} x_k)_{0 \leq j \leq n}$, there is a uniform factorization

$$\sum_{k=0}^{m} a_{jk} x_k = a y_j b, \quad 0 \leq j \leq n$$

with $a, y_j, b \in \mathcal{M}$ such that

$$\|D^{1/2p} a\|_{2p} \sup_j \|y_j\|_\infty \|b D^{1/2p}\|_{2p} \leq C \sup_{|\varepsilon_k|=1} \Big\| \sum_{k=0}^{m} \varepsilon_k x_k \Big\|_p,$$

where $C > 0$ is a uniform constant. Moreover, a similar result holds in the row and column case if we assume $p \geq 2$, and in the row+column case whenever $p \leq 2$ and $q = \infty$.

Second, we are going to prove a (p,q)-version of the preceding maximal inequality which corresponds to Theorem 14 (commutative case), 18 (tracial case), and 25 (nontracial case).

Theorem 34. *Let $A = (a_{jk})$ be a (p,q)-maximizing matrix. Then for each n and each choice of finitely many $\alpha_0, \ldots, \alpha_m \in \mathbb{C}$ and $x_0, \ldots, x_m \in \mathcal{M}$, i.e. for each choice of finitely many linear means $(\sum_{k=0}^{m} a_{jk} \alpha_k x_k)_{0 \leq j \leq n}$, there is a uniform factorization*

$$\sum_{k=0}^{m} a_{jk}\alpha_k x_k = a y_j b, \quad 0 \leq j \leq n$$

with $a, y_j, b \in \mathscr{M}$ such that

$$\|D^{1/2p}a\|_{2p} \sup_j \|y_j\|_\infty \|bD^{1/2p}\|_{2p} \leq C\Big(\sum_{k=0}^{m} |\alpha_k|^q\Big)^{1/q} \sup_{\|y'\|\leq 1} \Big(\sum_{k=0}^{m} |y'(D^{1/2p}x_kD^{1/2p})|^{q'}\Big)^{1/q'},$$

where $C > 0$ is a uniform constant. Moreover, a similar result holds in the column case if we assume $p \geq 2$.

Proof (of Theorems 33 and 34). Again we only check the symmetric case since the arguments for the column and row case are similar. For the proof of Theorem 33 note first that in view of the definition of the norm $\||\cdot\||_{p,n}^{\mathrm{sym}}$, we have to show that, given a $(2,2)$-maximizing matrix $A = (a_{jk})$, for each n and each choice of finitely many $x_0, \ldots, x_m \in \mathscr{M}$

$$\Big\|\Big|\Big(\sum_{k=0}^{m} a_{jk}x_k\Big)_{0\leq j\leq n}\Big\|\Big|_{p,n}^{\mathrm{sym}} \leq C \sup_{\|(\alpha_k)\|_\infty \leq 1} \Big\|\sum_{k=0}^{\infty} \alpha_k y_k\Big\|_p, \tag{3.70}$$

$C > 0$ some constant. Indeed, by Theorem 24 for each unconditionally convergent series $\sum_k y_k$ in $L_p(\mathscr{M}, \varphi)$ we have

$$\Big(\sum_{k=0}^{\infty} a_{jk}y_k\Big)_j \in L_p(\mathscr{M})[\ell_\infty].$$

If we apply a closed graph argument, this means that we find some constant $C > 0$ such that for every unconditionally convergent series $\sum_k y_k$ in $L_p(\mathscr{M}, \varphi)$

$$\Big\|\Big(\sum_{k=0}^{\infty} a_{jk}y_k\Big)_j\Big\|_{L_p(\mathscr{M})[\ell_\infty]} \leq C \sup_{\|(\alpha_k)\|_\infty \leq 1} \Big\|\sum_{k=0}^{\infty} \alpha_k y_k\Big\|_p.$$

In particular, we get for each n and each choice of finitely many $x_0, \ldots, x_m \in \mathscr{M}$ (i.e. for each choice of finitely many linear means $(\sum_{k=1}^{m} a_{jk}x_k)_{0\leq j\leq n}$),

$$\Big\|\Big(D^{1/2p}\Big(\sum_{k=0}^{m} a_{jk}x_k\Big)D^{1/2p}\Big)_{0\leq j\leq n}\Big\|_{L_p(\mathscr{M})[\ell_\infty^n]} \leq C \sup_{\|(\alpha_k)\|_\infty \leq 1} \Big\|\sum_{k=0}^{m} \alpha_k x_k\Big\|_p.$$

But from Proposition 11 we know that

$$\Big\|\Big(D^{1/2p}\Big(\sum_{k=0}^{m} a_{jk}x_k\Big)D^{1/2p}\Big)_{0\leq j\leq n}\Big\|_{L_p(\mathscr{M})[\ell_\infty^n]} = \Big\|\Big|\Big(\sum_{k=0}^{m} a_{jk}x_k\Big)_{0\leq j\leq n}\Big\|\Big|_{p,n}^{\mathrm{sym}},$$

which completes the proof of (3.70). Finally, it remains to prove Theorem 34. Here (in the symetric case) we have to show that for any (p,q)-maximizing matrix $A = (a_{jk})$, for each n and each choice of finitely many $x_0, \ldots, x_m \in \mathcal{M}$

$$\left\| \left(\sum_{k=0}^{m} a_{jk}x_k \right)_{0 \le j \le n} \right\|_{p,n}^{\text{sym}}$$
$$\le c \left(\sum_{k=0}^{m} |\alpha_k|^q \right)^{1/q} w_{q'} \left(D^{1/2p} x_k D^{1/2p}, L_p(\mathcal{M}) \right). \tag{3.71}$$

But of course we can argue exactly as in the proof of (3.70) replacing Theorem 24 by Theorem 25.   □

Clearly, Theorems 33 and 34 applied to matrices $A = S\Sigma D_{(1/\omega_k)}$, where S is a concrete summation matrix and (ω_k) its Weyl sequence, leads to various noncommutative extensions of classical inequalities for concrete summation methods – this time entirely formulated in terms of the integration space $(\mathcal{M}, \varphi)$ itself. Finally, it remains to prove Corollary 21.

Proof (of Corollary 21). From (2.34) we know that the matrix $A = \Sigma D_{(1/\log k)}$ defined by

$$a_{jk} = \begin{cases} \dfrac{1}{\log k} & k \le j \\ 0 & k > j \end{cases}$$

is $(2,2)$-maximizing. Hence for each choice of finitely many φ-orthonormal operators $x_0, \ldots, x_n \in \mathcal{M}$ and each choice of scalars $\alpha_0, \ldots \alpha_n$ we obtain from Theorem 34 (or (3.71))

$$\left\| \left(\sum_{k=0}^{j} \frac{\alpha_k x_k}{\log k} \right)_{0 \le j \le n} \right\|_{2,n}^{\text{col}} \le c \left(\sum_{k=0}^{n} |\alpha_k|^2 \right)^{1/2} w_2 \left(x_k, L_2(\mathcal{M}) \right).$$

But since the x_k's are φ-orthonormal, we know that $w_2(x_k, L_2(\mathcal{M})) = 1$ which gives the conclusion.   □

References

1. Alexits, G.: Convergence problems of orthogonal series. Translated from the German by I. Földer. International Series of Monographs in Pure and Applied Mathematics, Vol. 20. Pergamon Press, New York (1961)
2. Bennett, G.: Unconditional convergence and almost everywhere convergence. Z. Wahrscheinlichkeitstheorie und Verw. Gebiete **34**(2), 135–155 (1976)
3. Bennett, G.: Schur multipliers. Duke Math. J. **44**(3), 603–639 (1977)
4. Bergh, J., Löfström, J.: Interpolation spaces. An introduction. Springer-Verlag, Berlin (1976). Grundlehren der Mathematischen Wissenschaften, No. 223
5. Ciach, L.J.: Regularity and almost sure convergence. Ann. Univ. Mariae Curie-Skłodowska Sect. A **49**, 35–40 (1995)
6. Defant, A., Floret, K.: Tensor norms and operator ideals, *North-Holland Mathematics Studies*, vol. 176. North-Holland Publishing Co., Amsterdam (1993)
7. Defant, A., Junge, M.: Maximal theorems of Menchoff-Rademacher type in non-commutative L_q-spaces. J. Funct. Anal. **206**(2), 322–355 (2004). DOI 10.1016/j.jfa.2002.07.001. URL http://dx.doi.org/10.1016/j.jfa.2002.07.001
8. Defant, A., Mastyło, M., Michels, C.: Summing inclusion maps between symmetric sequence spaces, a survey. In: Recent progress in functional analysis (Valencia, 2000), *North-Holland Math. Stud.*, vol. 189, pp. 43–60. North-Holland, Amsterdam (2001)
9. Diestel, J., Jarchow, H., Tonge, A.: Absolutely summing operators, *Cambridge Studies in Advanced Mathematics*, vol. 43. Cambridge University Press, Cambridge (1995). 10.1017/CBO9780511526138. URL http://dx.doi.org/10.1017/CBO9780511526138
10. Dixmier, J.: Formes linéaires sur un anneau d'opérateurs. Bull. Soc. Math. France **81**, 9–39 (1953)
11. Dodds, P.G., Dodds, T.K., de Pagter, B.: Weakly compact subsets of symmetric operator spaces. Math. Proc. Cambridge Philos. Soc. **110**(1), 169–182 (1991). DOI 10.1017/S0305004100070225. URL http://dx.doi.org/10.1017/S0305004100070225
12. Dodds, P.G., Dodds, T.K., de Pagter, B.: Fully symmetric operator spaces. Integral Equations Operator Theory **15**(6), 942–972 (1992). DOI 10.1007/BF01203122. URL http://dx.doi.org/10.1007/BF01203122
13. Dodds, P.G., Dodds, T.K.Y., de Pagter, B.: Noncommutative Banach function spaces. Math. Z. **201**(4), 583–597 (1989). DOI 10.1007/BF01215160. URL http://dx.doi.org/10.1007/BF01215160
14. Dodds, P.G., Dodds, T.K.Y., de Pagter, B.: Noncommutative Köthe duality. Trans. Amer. Math. Soc. **339**(2), 717–750 (1993). DOI 10.2307/2154295. URL http://dx.doi.org/10.2307/2154295

15. Erdös, P.: On the convergence of trigonometric series. J. Math. Phys. Mass. Inst. Tech. **22**, 37–39 (1943)
16. Fack, T.: Type and cotype inequalities for noncommutative L^p-spaces. J. Operator Theory **17**(2), 255–279 (1987)
17. Fack, T., Kosaki, H.: Generalized s-numbers of τ-measurable operators. Pacific J. Math. **123**(2), 269–300 (1986). URL http://projecteuclid.org/getRecord?id=euclid.pjm/1102701004
18. Garling, D.J.H., Tomczak-Jaegermann, N.: The cotype and uniform convexity of unitary ideals. Israel J. Math. **45**(2-3), 175–197 (1983). DOI 10.1007/BF02774015. URL http://dx.doi.org/10.1007/BF02774015
19. Goldstein, S.: Theorems on almost everywhere convergence in von Neumann algebras. J. Operator Theory **6**(2), 233–311 (1981)
20. Goldstein, S., Łuczak, A.: On the Rademacher-Menshov theorem in von Neumann algebras. Studia Sci. Math. Hungar. **22**(1-4), 365–377 (1987)
21. Grothendieck, A.: Résumé de la théorie métrique des produits tensoriels topologiques. Bol. Soc. Mat. São Paulo **8**, 1–79 (1956)
22. Grothendieck, A.: Réarrangements de fonctions et inégalités de convexité dans les algèbres de von Neumann munies d'une trace. In: Séminaire Bourbaki, Vol. 3 (1954–1956), pp. Exp. No. 113, 127–139. Soc. Math. France, Paris (1995)
23. Haagerup, U.: Normal weights on W^*-algebras. J. Functional Analysis **19**, 302–317 (1975)
24. Haagerup, U.: L^p-spaces associated with an arbitrary von Neumann algebra. In: Algèbres d'opérateurs et leurs applications en physique mathématique (Proc. Colloq., Marseille, 1977), *Colloq. Internat. CNRS*, vol. 274, pp. 175–184. CNRS, Paris (1979)
25. Haagerup, U.: Decompositions of completely bounded maps on operator algebras. Unpublished manuscript (Sept. 1980)
26. Haagerup, U., Junge, M., Xu, Q.: A reduction method for noncommutative L_p-spaces and applications. Trans. Amer. Math. Soc. **362**(4), 2125–2165 (2010). DOI 10.1090/S0002-9947-09-04935-6. URL http://dx.doi.org/10.1090/S0002-9947-09-04935-6
27. Hensz, E.: On a Weyl theorem in von Neumann algebras. Bull. Polish Acad. Sci. Math. **35**(3-4), 195–201 (1987)
28. Hensz, E.: Orthogonal series and strong laws of large numbers in von Neumann algebras. In: Quantum probability and applications, IV (Rome, 1987), *Lecture Notes in Math.*, vol. 1396, pp. 221–228. Springer, Berlin (1989). DOI 10.1007/BFb0083554. URL http://dx.doi.org/10.1007/BFb0083554
29. Hensz, E.: Strong laws of large numbers for orthogonal sequences in von Neumann algebras. In: Probability theory on vector spaces, IV (Łańcut, 1987), *Lecture Notes in Math.*, vol. 1391, pp. 112–124. Springer, Berlin (1989). DOI 10.1007/BFb0083385. URL http://dx.doi.org/10.1007/BFb0083385
30. Hensz, E.: The Cesàro means and strong laws of large numbers for orthogonal sequences in von Neumann algebras. Proc. Amer. Math. Soc. **110**(4), 939–945 (1990). DOI 10.2307/2047740. URL http://dx.doi.org/10.2307/2047740
31. Hensz, E.: Strong laws of large numbers for nearly orthogonal sequences of operators in von Neumann algebras. In: Proceedings of the Second Winter School on Measure Theory (Lipovský Ján, 1990), pp. 85–91. Jedn. Slovensk. Mat. Fyz., Bratislava (1990)
32. Hensz, E.: Szeregi i mocne prawa wielkich liczb w przestrzeni l_2 nad algebra von neumanna. Acta Universitatis Lodziensis, Wydawnictwo Uniwerytetu Lodzkiego, Lodz (1992)
33. Hensz, E., Jajte, R.: Pointwise convergence theorems in L_2 over a von Neumann algebra. Math. Z. **193**(3), 413–429 (1986). DOI 10.1007/BF01229809. URL http://dx.doi.org/10.1007/BF01229809
34. Hensz, E., Jajte, R., Paszkiewicz, A.: The unconditional pointwise convergence of orthogonal series in L_2 over a von Neumann algebra. Colloq. Math. **69**(2), 167–178 (1995)
35. Ignaczak, E.: Convergence of subsequences in W^*-algebras. Demonstratio Math. **25**(1-2), 305–316 (1992)

36. Jajte, R.: Strong limit theorems for orthogonal sequences in von Neumann algebras. Proc. Amer. Math. Soc. **94**(2), 229–235 (1985). DOI 10.2307/2045381. URL http://dx.doi.org/ 10.2307/2045381

37. Jajte, R.: Strong limit theorems in noncommutative probability, *Lecture Notes in Mathematics*, vol. 1110. Springer-Verlag, Berlin (1985)

38. Jajte, R.: Strong limit theorems in noncommutative L_2-spaces, *Lecture Notes in Mathematics*, vol. 1477. Springer-Verlag, Berlin (1991)

39. Junge, M.: Doob's inequality for non-commutative martingales. J. Reine Angew. Math. **549**, 149–190 (2002). DOI 10.1515/crll.2002.061. URL http://dx.doi.org/10.1515/ crll.2002.061

40. Junge, M., Sherman, D.: Noncommutative L^p modules. J. Operator Theory **53**(1), 3–34 (2005)

41. Junge, M., Xu, Q.: Noncommutative Burkholder/Rosenthal inequalities. Ann. Probab. **31**(2), 948–995 (2003). DOI 10.1214/aop/1048516542. URL http://dx.doi.org/10.1214/ aop/1048516542

42. Junge, M., Xu, Q.: Noncommutative maximal ergodic theorems. J. Amer. Math. Soc. **20**(2), 385–439 (2007). DOI 10.1090/S0894-0347-06-00533-9. URL http://dx.doi.org/ 10.1090/S0894-0347-06-00533-9

43. Kaczmarz, S.: Über die Summierbarkeit der Orthogonalreihen. Math. Z. **26**(1), 99–105 (1927). DOI 10.1007/BF01475443. URL http://dx.doi.org/10.1007/BF01475443

44. Kadison, R.V., Ringrose, J.R.: Fundamentals of the theory of operator algebras. Vol. I,II, *Graduate Studies in Mathematics*, vol. 15,16. American Mathematical Society, Providence, RI (1997). Corrected reprint of the 1983/1986 originals

45. Kalton, N.J., Sukochev, F.A.: Symmetric norms and spaces of operators. J. Reine Angew. Math. **621**, 81–121 (2008). DOI 10.1515/CRELLE.2008.059. URL http://dx.doi.org/ 10.1515/CRELLE.2008.059

46. Kantorovič, L.: Some theorems on the almost everywhere convergence. Dokl. Akad. Nauk USSR **14**, 537–540 (1937)

47. Kashin, B.S., Saakyan, A.A.: Orthogonal series, *Translations of Mathematical Monographs*, vol. 75. American Mathematical Society, Providence, RI (1989). Translated from the Russian by Ralph P. Boas, Translation edited by Ben Silver

48. Kosaki, H.: Applications of the complex interpolation method to a von Neumann algebra: noncommutative L^p-spaces. J. Funct. Anal. **56**(1), 29–78 (1984). DOI 10.1016/0022-1236(84) 90025-9. URL http://dx.doi.org/10.1016/0022-1236(84)90025-9

49. Kunze, R.A.: L_p Fourier transforms on locally compact unimodular groups. Trans. Amer. Math. Soc. **89**, 519–540 (1958)

50. Kwapień, S., Pełczyński, A.: The main triangle projection in matrix spaces and its applications. Studia Math. **34**, 43–68 (1970)

51. Lance, E.C.: Ergodic theorems for convex sets and operator algebras. Invent. Math. **37**(3), 201–214 (1976)

52. Lance, E.C.: Martingale convergence in von Neumann algebras. Math. Proc. Cambridge Philos. Soc. **84**(1), 47–56 (1978). DOI 10.1017/S0305004100054864. URL http://dx.doi.org/10.1017/S0305004100054864

53. Lindenstrauss, J., Tzafriri, L.: Classical Banach spaces. I. Springer-Verlag, Berlin (1977). Sequence spaces, Ergebnisse der Mathematik und ihrer Grenzgebiete, Vol. 92

54. Lindenstrauss, J., Tzafriri, L.: Classical Banach spaces. II, *Ergebnisse der Mathematik und ihrer Grenzgebiete [Results in Mathematics and Related Areas]*, vol. 97. Springer-Verlag, Berlin (1979). Function spaces

55. Lust-Piquard, F.: A Grothendieck factorization theorem on 2-convex Schatten spaces. Israel J. Math. **79**(2-3), 331–365 (1992). DOI 10.1007/BF02808225. URL http://dx.doi.org/ 10.1007/BF02808225

56. Lust-Piquard, F., Xu, Q.: The little Grothendieck theorem and Khintchine inequalities for symmetric spaces of measurable operators. J. Funct. Anal. **244**(2), 488–503 (2007). DOI 10.1016/j.jfa.2006.09.003. URL http://dx.doi.org/10.1016/j.jfa.2006. 09.003

57. Mastyło, M.: Factorization of operators through abstract lorentz interpolation spaces. preprint

58. Maurey, B.: Théorèmes de factorisation pour les opérateurs linéaires à valeurs dans les espaces L^p. Société Mathématique de France, Paris (1974). With an English summary, Astérisque, No. 11

59. Maurey, B., Nahoum, A.: Applications radonifiantes dans l'espace des séries convergentes. C. R. Acad. Sci. Paris Sér. A-B **276**, A751–A754 (1973)

60. Menchoff, D.: Sur les séries de fonctions orthogonales. (Première Partie. La convergence.). Fundamenta Math. **4**, 82–105 (1923)

61. Menchoff, D.: Sur la sommation des séries de fonctions orthogonales. Comptes Rendus Acad. Sci. Paris **180**, 2011–2013 (1925)

62. Menchoff, D.: Sur les séries de fonctions orthogonales. ii. Fundamenta Math. **8**, 56–108 (1926)

63. Móricz, F.: On the Cesàro means of orthogonal sequences of random variables. Ann. Probab. **11**(3), 827–832 (1983). URL http://links.jstor.org/sici?sici=0091-1798(198308)11:3<827:OTCMOO>2.0.CO;2-Y&origin=MSN

64. Móricz, F.: SLLN and convergence rates for nearly orthogonal sequences of random variables. Proc. Amer. Math. Soc. **95**(2), 287–294 (1985). DOI 10.2307/2044529. URL http://dx.doi.org/10.2307/2044529

65. Murphy, G.J.: C^*-algebras and operator theory. Academic Press Inc., Boston, MA (1990)

66. Nelson, E.: Notes on non-commutative integration. J. Functional Analysis **15**, 103–116 (1974)

67. Orlicz, W.: Zur theorie der orthogonalreihen. Bulletin Acad. Polonaise Sci. Lettres (Cracovie) (A) **1927**, 81–115 (1927)

68. Ørno, P.: A note on unconditionally converging series in L_p. Proc. Amer. Math. Soc. **59**(2), 252–254 (1976)

69. Ovčinnikov, V.I.: Symmetric spaces of measurable operators. Dokl. Akad. Nauk SSSR **191**, 769–771 (1970)

70. Ovčinnikov, V.I.: Symmetric spaces of measurable operators. Voronež. Gos. Univ. Trudy Naučn.-Issled. Inst. Mat. VGU (3), 88–107 (1971)

71. Paszkiewicz, A.: Convergence almost everywhere in W^*-algebras. In: Quantum probability and applications, II (Heidelberg, 1984), Lecture Notes in Math., vol. 1136, pp. 420–427. Springer, Berlin (1985). DOI 10.1007/BFb0074490. URL http://dx.doi.org/10.1007/BFb0074490

72. Paszkiewicz, A.: Convergences in W^*-algebras. J. Funct. Anal. **69**(2), 143–154 (1986). DOI 10.1016/0022-1236(86)90086-8. URL http://dx.doi.org/10.1016/0022-1236(86)90086-8

73. Paszkiewicz, A.: Convergences in W^*-algebras—their strange behaviour and tools for their investigation. In: Quantum probability and applications, IV (Rome, 1987), Lecture Notes in Math., vol. 1396, pp. 279–294. Springer, Berlin (1989). DOI 10.1007/BFb0083558. URL http://dx.doi.org/10.1007/BFb0083558

74. Paszkiewicz, A.: A limit in probability in a W^*-algebra is unique. J. Funct. Anal. **90**(2), 429–444 (1990). DOI 10.1016/0022-1236(90)90091-X. URL http://dx.doi.org/10.1016/0022-1236(90)90091-X

75. Petz, D.: Quasi-uniform ergodic theorems in von Neumann algebras. Bull. London Math. Soc. **16**(2), 151–156 (1984). DOI 10.1112/blms/16.2.151. URL http://dx.doi.org/10.1112/blms/16.2.151

76. Pietsch, A.: Operator ideals, North-Holland Mathematical Library, vol. 20. North-Holland Publishing Co., Amsterdam (1980). Translated from German by the author

77. Pisier, G.: Factorization of linear operators and geometry of Banach spaces, CBMS Regional Conference Series in Mathematics, vol. 60. Published for the Conference Board of the Mathematical Sciences, Washington, DC (1986)

78. Pisier, G.: Similarity problems and completely bounded maps, Lecture Notes in Mathematics, vol. 1618. Springer-Verlag, Berlin (1996)

79. Pisier, G.: Non-commutative vector valued L_p-spaces and completely p-summing maps. Astérisque (247), vi+131 (1998)

80. Pisier, G., Xu, Q.: Non-commutative L^p-spaces. In: Handbook of the geometry of Banach spaces, Vol. 2, pp. 1459–1517. North-Holland, Amsterdam (2003). DOI 10.1016/ S1874-5849(03)80041-4. URL http://dx.doi.org/10.1016/S1874-5849(03) 80041-4

81. Rademacher, H.: Einige Sätze über Reihen von allgemeinen Orthogonalfunktionen. Math. Ann. **87**(1-2), 112–138 (1922). DOI 10.1007/BF01458040. URL http://dx.doi.org/ 10.1007/BF01458040

82. Révész, P.: The laws of large numbers. Probability and Mathematical Statistics, Vol. 4. Academic Press, New York (1968)

83. Schwartz, L.: Mesures cylindriques et applications radonifiantes dans les espaces de suites. In: Proc. Internat. Conf. on Functional Analysis and Related Topics (Tokyo, 1969), pp. 41–59. University of Tokyo Press, Tokyo (1970)

84. Segal, I.E.: A non-commutative extension of abstract integration. Ann. of Math. (2) **57**, 401– 457 (1953)

85. Sukochev, F.A.: (en)-invariant properties of symmetric spaces of measurable operators. Dokl. Akad. Nauk UzSSR (7), 6–8 (1985)

86. Sukochev, F.A.: Order properties of norms of symmetric spaces of measurable operators. In: Mathematical analysis and probability theory, pp. 49–54, 94. Tashkent. Gos. Univ., Tashkent (1985)

87. Sukochev, F.A.: Construction of noncommutative symmetric spaces. Dokl. Akad. Nauk UzSSR (8), 4–6 (1986)

88. Sukochev, F.A., Chilin, V.I.: Symmetric spaces over semifinite von Neumann algebras. Dokl. Akad. Nauk SSSR **313**(4), 811–815 (1990)

89. Takesaki, M.: Theory of operator algebras. I,II,III, *Encyclopaedia of Mathematical Sciences*, vol. 124,125,127. Springer-Verlag, Berlin (2002/2003/2003). Reprint of the first (1979) edition, Operator Algebras and Non-commutative Geometry, 5,6,8

90. Terp, M.: l_p-spaces associated with von neumann algebras. preprint (1981)

91. Toeplitz, O.: Über allgemeine lineare Mittelbildungen. Prace mat.-fiz. **22**, 113–119 (1913)

92. Tomczak-Jaegermann, N.: Banach-Mazur distances and finite-dimensional operator ideals, *Pitman Monographs and Surveys in Pure and Applied Mathematics*, vol. 38. Longman Scientific & Technical, Harlow (1989)

93. Weidmann, J.: Linear operators in Hilbert spaces, *Graduate Texts in Mathematics*, vol. 68. Springer-Verlag, New York (1980). Translated from the German by Joseph Szücs

94. Wojtaszczyk, P.: Banach spaces for analysts, *Cambridge Studies in Advanced Mathematics*, vol. 25. Cambridge University Press, Cambridge (1991). DOI 10.1017/CBO9780511608735. URL http://dx.doi.org/10.1017/CBO9780511608735

95. Xu, Q.: Convexité uniforme des espaces symétriques d'opérateurs mesurables. C. R. Acad. Sci. Paris Sér. I Math. **309**(5), 251–254 (1989)

96. Xu, Q.H.: Analytic functions with values in lattices and symmetric spaces of measurable operators. Math. Proc. Cambridge Philos. Soc. **109**(3), 541–563 (1991). DOI 10.1017/ S030500410006998X. URL http://dx.doi.org/10.1017/S030500410006998X

97. Zygmund, A.: Sur la sommation des séries de fonctions orthogonales. Bulletin Acad. Polonaise Sci. Lettres (Cracovie) (A) **1927**, 295–308 (1927)

98. Zygmund, A.: Trigonometric series. Vol. I, II, third edn. Cambridge Mathematical Library. Cambridge University Press, Cambridge (2002). With a foreword by Robert A. Fefferman

Symbols

Author Index

Subject Index

LECTURE NOTES IN MATHEMATICS  **Springer**

Edited by J.-M. Morel, F. Takens, B. Teissier, P.K. Maini

Editorial Policy (for the publication of monographs)

1. Lecture Notes aim to report new developments in all areas of mathematics and their applications - quickly, informally and at a high level. Mathematical texts analysing new developments in modelling and numerical simulation are welcome.

    Monograph manuscripts should be reasonably self-contained and rounded off. Thus they may, and often will, present not only results of the author but also related work by other people. They may be based on specialised lecture courses. Furthermore, the manuscripts should provide sufficient motivation, examples and applications. This clearly distinguishes Lecture Notes from journal articles or technical reports which normally are very concise. Articles intended for a journal but too long to be accepted by most journals, usually do not have this "lecture notes" character. For similar reasons it is unusual for doctoral theses to be accepted for the Lecture Notes series, though habilitation theses may be appropriate.

2. Manuscripts should be submitted either online at www.editorialmanager.com/lnm to Springer's mathematics editorial in Heidelberg, or to one of the series editors. In general, manuscripts will be sent out to 2 external referees for evaluation. If a decision cannot yet be reached on the basis of the first 2 reports, further referees may be contacted: The author will be informed of this. A final decision to publish can be made only on the basis of the complete manuscript, however a refereeing process leading to a preliminary decision can be based on a pre-final or incomplete manuscript. The strict minimum amount of material that will be considered should include a detailed outline describing the planned contents of each chapter, a bibliography and several sample chapters.

    Authors should be aware that incomplete or insufficiently close to final manuscripts almost always result in longer refereeing times and nevertheless unclear referees' recommendations, making further refereeing of a final draft necessary.

    Authors should also be aware that parallel submission of their manuscript to another publisher while under consideration for LNM will in general lead to immediate rejection.

3. Manuscripts should in general be submitted in English. Final manuscripts should contain at least 100 pages of mathematical text and should always include

    – a table of contents;
    – an informative introduction, with adequate motivation and perhaps some historical remarks: it should be accessible to a reader not intimately familiar with the topic treated;
    – a subject index: as a rule this is genuinely helpful for the reader.

    For evaluation purposes, manuscripts may be submitted in print or electronic form (print form is still preferred by most referees), in the latter case preferably as pdf- or zipped ps-files. Lecture Notes volumes are, as a rule, printed digitally from the authors' files. To ensure best results, authors are asked to use the LaTeX2e style files available from Springer's web-server at:

    ftp://ftp.springer.de/pub/tex/latex/svmonot1/ (for monographs) and
    ftp://ftp.springer.de/pub/tex/latex/svmultt1/ (for summer schools/tutorials).

Additional technical instructions, if necessary, are available on request from: lnm@springer.com.

4. Careful preparation of the manuscripts will help keep production time short besides ensuring satisfactory appearance of the finished book in print and online. After acceptance of the manuscript authors will be asked to prepare the final LaTeX source files and also the corresponding dvi-, pdf- or zipped ps-file. The LaTeX source files are essential for producing the full-text online version of the book (see
http://www.springerlink.com/openurl.asp?genre=journal&issn=0075-8434 for the existing online volumes of LNM).
   The actual production of a Lecture Notes volume takes approximately 12 weeks.

5. Authors receive a total of 50 free copies of their volume, but no royalties. They are entitled to a discount of 33.3% on the price of Springer books purchased for their personal use, if ordering directly from Springer.

6. Commitment to publish is made by letter of intent rather than by signing a formal contract. Springer-Verlag secures the copyright for each volume. Authors are free to reuse material contained in their LNM volumes in later publications: a brief written (or e-mail) request for formal permission is sufficient.

Addresses:
Professor J.-M. Morel, CMLA,
École Normale Supérieure de Cachan,
61 Avenue du Président Wilson, 94235 Cachan Cedex, France
E-mail: Jean-Michel.Morel@cmla.ens-cachan.fr

Professor F. Takens, Mathematisch Instituut,
Rijksuniversiteit Groningen, Postbus 800,
9700 AV Groningen, The Netherlands
E-mail: F.Takens@rug.nl

Professor B. Teissier, Institut Mathématique de Jussieu,
UMR 7586 du CNRS, Équipe "Géométrie et Dynamique",
175 rue du Chevaleret,
75013 Paris, France
E-mail: teissier@math.jussieu.fr

For the "Mathematical Biosciences Subseries" of LNM:

Professor P.K. Maini, Center for Mathematical Biology,
Mathematical Institute, 24-29 St Giles,
Oxford OX1 3LP, UK
E-mail: maini@maths.ox.ac.uk

Springer, Mathematics Editorial, Tiergartenstr. 17,
69121 Heidelberg, Germany,
Tel.: +49 (6221) 487-259
Fax: +49 (6221) 4876-8259
E-mail: lnm@springer.com